D1196807

"More than any time in human history, we have access to mountains of data about ourselves. *Hacking H(app)iness* is the first book to show us how to leverage this information as a path to happiness, rather than a source of misery."

—**Adam Grant**, *New York Times*–bestselling
author of *Give and Take*, and Wharton professor

"In *Hacking H(app)iness*, John C. Havens makes the persuasive case that a key to happiness in the digital age is being able to control and leverage your personal data for your own benefit. It's a must-read for anyone who wants to better understand the interplay of economics, innovation, and the rising personal data sector, and how you can make better, smarter decisions when you're in charge of your own data."

—**Shane Green**, co-founder and CEO of Personal

"I've met and spoken with literally hundreds of people about aging and the consequences of isolation. Most of them knew the space; many of them understood the emotional impacts, but only John *felt* it. He intuitively understood how our societal focus on physical health was obscuring our view on emotional health."

—**Iggy Fanlo**, co-founder and CEO of Lively

"The unexamined digital life is walking along an unstable ledge of happiness, in an era of digital exuberance. John C. Havens's *Hacking H(app)iness* is the balancing stick that allows us to synthesize and leverage technology by understanding the evolutionary value of one's digital blueprint, so that well-being and happiness can emerge."

—**Judy Martin**, founder of WorkLifeNation.com
and contributor for *Forbes* and NPR

"John C. Havens gives us an illuminating examination of how emerging technology can be harnessed to promote individual, community, corporate, and global happiness. As one who studies intrinsic motivation, achievement, and happiness, I enjoyed John's rare emphasis on altruistically serving others as a path toward greater happiness and health."

—**John Mark Froiland, Ph.D.**, assistant professor
of psychology, University of Northern Colorado

"In the twentieth century, we made great progress in terms of our material wealth, but we're not really any happier. In this insightful book, John Havens shows us how the new century will bring us opportunities to improve our general well-being. Rather than keeping up with the Joneses, he explains how we can use technology to actually improve our lives. It is a truly remarkable work."

—**Greg Satell**, contributing writer for *Forbes*

"John Havens has written a comprehensive guide through our complicated digital lives, carefully examining the benefits of the data-driven pursuit of happiness through the lens of an enlightened idealist. A must-read for anyone interested in a humane future of connectivity."   —**Tim Leberecht**, chief marketing officer of NBBJ

"John Havens is leading the charge to change the way we talk and think about digital consumer technology. Rather than simply asking whether the latest gadget is faster or has more features, John encourages us to ask such questions as 'Will this make me happier?' And it's not just a rhetorical ploy; he wants us to think through the question sincerely. John avoids the knee-jerk conclusions of both the techno-fanboy and neo-Luddite camps—to the occasional irritation of both—which makes his work all the more important."

—**Brian Wassom**, augmented reality
law expert, partner at Honigman Miller
Schwartz and Cohn LLP

"This book shows us that happiness can be an active pursuit—a journey filled with data and optimization, with satisfaction as the ultimate goal. Just reading this book made me happier."

—**Ari Meisel**, author of *Less Doing,
More Living*, founder of the
Art of Less Doing, and Ironman triathlete

"*Hacking H(app)iness* covers a whole range of technologies that are all emerging and looks at them from a positive perspective to see how they can help people, our communities, and the world. John's approach is refreshing and adds new perspectives to consider how we as a society make considerations about what technologies to adopt and how they might fit together for the benefit of the whole."

—**Kaliya**, aka "Identity Woman"

"In *Hacking H(app)iness*, John C. Havens proves the importance of measuring our lives to identify our purpose versus just increasing profits or productivity. By showing how altruistic actions can increase happiness, Havens also provides a road map to scaling (or hacking) how the world perceives value, where currency will be based on compassion versus capitalism."

—**Aaron Hurst**, author of *The Purpose
Economy*, and CEO of Imperative

Hacking H(app)iness

Hacking H(app)iness

WHY YOUR PERSONAL DATA COUNTS
AND HOW TRACKING IT CAN
CHANGE THE WORLD

[JOHN C. HAVENS]

JEREMY P. TARCHER/PENGUIN
a member of Penguin Group (USA)
New York

JEREMY P. TARCHER/PENGUIN
Published by the Penguin Group
Penguin Group (USA) LLC
375 Hudson Street
New York, New York 10014

USA · Canada · UK · Ireland · Australia
New Zealand · India · South Africa · China

penguin.com
A Penguin Random House Company

Most Tarcher/Penguin books are available at special quantity discounts for bulk purchase
for sales promotions, premiums, fund-raising, and educational needs. Special books or
book excerpts also can be created to fit specific needs. For details, write:
Special.Markets@us.penguingroup.com.

Library of Congress Cataloging-in-Publication Data

Havens, John C.
Hacking happiness : why your personal data counts and how tracking it can change
the world / John C. Havens.
p.   cm.
ISBN 978-0-399-16531-3
1. Technological innovations—Social aspects.   2. Self-monitoring.
3. Data mining—Social aspects.   4. Well-being.   5. Happiness.   I. Title.
HM846.H38   2014   2013038876
303.48'3—dc23

Printed in the United States of America
1   3   5   7   9   10   8   6   4   2

BOOK DESIGN BY TANYA MAIBORODA

This book is dedicated to

David W. Havens, M.D. — the man who listened.

CONTENTS

INTRODUCTION

It's strange to look at a screen and see a number that represents your life.

My dad had died three months earlier and I was grieving in my own way. Like a lot of people dealing with loss, that way involved distraction. The number I was looking at was a score from a service called Klout, a self-described "authority for online influence." On Twitter, Facebook, and other social networks, a series of algorithms determined a number between one and one hundred, a representation of the digital me.

I don't remember my score. I just remember being wounded. I felt cheated that the number seemed low and someone I didn't know was controlling the way I was valued. I tried to pretend the number didn't bother me, but it did. I felt anxious, and began planning how I'd write a certain number of tweets or Facebook posts to game the system. I'd comment for the sake of increasing my influence, whether or not I really had anything to say.

Then I stopped. I wondered who I had become if I was scripting my life in such a way that I was shaping my insights to either fit into 140 characters or be pithy enough to play well on Facebook. I realized I was living my life in spurts long enough to get a good sound bite.

I thought of my dad. I thought of my kids. I thought of what I'd leave behind as a legacy, and I took a moment to reflect on my life, instead of commenting on it. This was a risk. I knew reflecting meant dealing with the raw truth of who I was, but I genuinely wanted to understand the measure of my life. Fortunately, my answer came fairly quickly, and from a very deep place—I wanted my life to count.

My perspective changed immediately. I felt an internal shift, where my desire to create influence was supplanted by a need to create impact. This realization transformed my anxiety into a sense of well-being. Creating impact meant I'd pursue actions that had potential for helping others, versus focusing on digital influence, where I'd always be seeking immediate attention. Focusing on impact also felt pragmatic. When I pursued influence, I always felt exactly as I did when I pursued happiness just for the sake of it—narcissistic and exhausted. Intrinsic joy for me had always come from pursuing actions where happiness came as a *result* of dealing with adversity or meaningful challenge.

And regarding the measure of my life: I realized I meant the phrase literally. In the realm of technology, where I thought I was an expert, other people were making decisions that would determine how I'd be valued in the digital world. And I realized that similar decisions about online or mobile behavior made by other organizations would start to aggregate around the idea of people's data. Data, I knew, that was already being sold in convoluted ways online, with people giving away their digital DNA in exchange for a onetime offer.

I saw a path toward an inevitable future where our digital identities were becoming tangible currency and our worth would be determined by algorithms. I saw that technologies like augmented reality would create an atmosphere where people would see the digital representations of other people before getting to know them in person. I saw that a tiny population of individuals determining digital rankings would literally alter how we would view the world, and how others would view us.

And that's when I got mad. I got deeply, deeply pissed.

I should get to determine my identity. *I* should get to determine how the digital journal of my life in the form of my data gets broadcast, sold, or valued.

And so should you if you want your life in the digital world to count.

As you may already have guessed, the happiness described in this book is not focused on mood but is the result of an introspective examination of what brings you purpose or meaning. In other words, you have to take action regarding what defines your life to truly see what brings you joy.

That's where the idea of "hacking" comes into play. The main definition of hacking involves cybercrime. This is not my focus. I'm referring to a different sort of "hacking" that involves the creative reimagining of a long-held idea for the sake of joyful discovery. That's the spirit in which I invite you to read this book. I see data and technology, when leveraged via informed choice, as being instrumental toward insightful living.

The idea of self-examination, focused on your digital as well as real self, is where the subtitle for the book comes from—*why your personal data counts and how tracking it can change the world*. Your data counts because it represents who you are, and has value intrinsically and economically. I'm sure you're aware that other people are already tracking your data—marketers, governments, and organizations of every size around the world. That's why a primary goal for me in writing this book is to simply help you do the same, so you can enjoy the benefits of managing your data to gain insights that can increase your well-being.

The (app) part of *Hacking H(app)iness* refers to the number of apps and other technologies I cover in the book that can measure or track your emotional and physical health and well-being. I've also broken up the book into three parts based on the (app) acronym, which I explain in detail later in the introduction. While this isn't a traditional self-help or how-to book, I have provided numerous examples and case studies of how technology applied to emotion and economics can improve your life and help define what brings you meaning.

Here's what I've set out to prove in this book:

- **Data is getting personal.** Data about people's sleep, dietary, and work habits, sex lives, and emotions is being collected online and analyzed. Our current online economy is built on the managing of people's data without their full knowledge of the process. It's a dangerous precedent that needs to be reversed.
- **Happiness can be quantified and increased.** The science of positive psychology is empirically based. Mobile sensors in our phones and data from the world around us can contribute information about our lives that we can utilize to increase positive well-being and essential, long-lasting happiness. While emotions are ephemeral and subjective by nature, data-identifying triggers leading to or around them are being leveraged to improve people's lives in revolutionary ways.
- **The happiness economy is redefining wealth.** Countries such as Bhutan, the United Kingdom, Brazil, China, and the United States are using happiness indicators that reflect multiple metrics beyond money to measure and improve the lives of their citizens. People are being encouraged to leverage skills and talents for

civic engagement that are providing previously untapped stores of resources that are changing the world for good.

Here's why you'll benefit from reading *Hacking H(app)iness*:

- **Informed Choice.** What you do with your data is your decision. But I want you to see vividly that complacency about giving away your digital identity is not a choice you should allow for yourself or loved ones any longer.
- **Joyful Discovery.** Measuring your life isn't easy, but with mobile tools and positive psychology, there's never been a better time to start the journey. There's not a one-size-fits-all formula for this process, as it's inherently your own, but your life will be richer by examining your value, and a great deal of this book is dedicated to showing you how.
- **The Currency of Connection.** Economics at its core is not about numbers or statistics as much as it's a way of measuring and expressing value. You care about what you count—where your treasure is reflects your heart. People invent economic ideals. You're allowed to evaluate what money and time mean for your life outside of established mind-sets. Soon, currency will revolve around your positive actions and reputation more than around wealth or words.

I'll explain these concepts more fully before describing the breakdown of the book.

Unpacking the Hacking

Your Identity in Data

Let me be clear about a critical aspect of your digital identity—your data is being sold and you're giving it away for free.

The model of tracking and behavioral targeting prevalent in the online world is accelerating the sale of your data, or your digital identity. While the companies involved in this form of commerce may not be mendacious in nature, this current advertising model erodes trust and is primed for disruption. If you're like most, you may be accepting of this model because you don't see all of its implications. Shane Green of Personal.com calls this phenomenon a form of the "Stockholm syndrome" in *Power-Curve Society: The Future of Innovation, Opportunity and Social Equity in the Emerging Networked Economy.*[1] The actual syndrome occurs when a victim identifies with his captor over time, fostering feelings of empathy and submission. In our current Internet economy, citizens willingly sign away data that's used in a variety of exploitative ways by government and business. The fact that they're agreeing to byzantine legal agreements not understandable by most attorneys has faded in importance for the perceived benefit of "freemium" services offered in exchange for previous personal information.

Your data has value. Your actions, in aggregate with other people like you, provide clues to behavior that are a form of market research. To give an example of how our personal data has a *monetary* value, Steffan Heuer and Pernille Tranberg provide the following example in their e-book *Fake It! Your Guide to Digital Self-Defense*:

> So how much are all of us worth online? The estimates vary, depending on whom you ask and what method you use to calculate. Lawyers in the U.S. representing users who felt their privacy was violated by apps that stole their address books from their mobile devices, estimated a price for each contact uploaded without permission. Their guess: between sixty cents and three dollars for each stolen contact, because services such as Path or Instagram can use them to acquire

new users and sell their information, or use it for targeted advertising down the road.[2]

Where you are comfortable revealing this data or sharing rights to access it, that's your call. In the near future, you'll have more choices on how to use your data, as Shane Green told me when I interviewed him on this subject:

In the current world, there's a shotgun approach toward monetizing data online. This practice will begin to shift toward the individual so they're not "selling their data" but will be "compensated for access to their data." This new world will be a place where the individual is empowered and becomes savvy about how data reaches them.[3]

But that time has not yet arrived. In December 2012, the Federal Trade Commission announced it would be studying the data broker industry's collection and use of consumer data, based on an earlier FTC report, *Protecting Consumer Privacy in an Era of Rapid Change: Recommendations for Businesses and Policymakers.*[4] The report noted that data brokers often don't interact directly with consumers, even though they're collecting and selling information about them. This means the average person isn't even aware data brokers exist or who may be buying and selling their data. As the FTC press release notes, "This lack of transparency also means that even when data brokers offer consumers the ability to access their data, or provide other tools, many consumers do not know how to exercise this right. *There are no current laws requiring data brokers to maintain the privacy of consumer data* unless they use that data for credit, employment, insurance, housing, or other similar purposes" (italics mine).[5]

Sadly, it's not just data brokers who are ignoring data privacy. The release of Facebook's Graph Search in March 2013 allows any

Facebook user to type in a name and see photos of other users who may not have wanted those photos to be public. As Sarah Perez notes in TechCrunch, Graph Search deepens confusion around privacy because other people's posting behavior affects you whether or not they've asked your permission.[6] As an example, she explains how a college friend had posted and tagged a slightly embarrassing picture of her without her knowledge or consent, and notes how complicated Facebook's privacy settings are to figure out, and how time-consuming it can be to remove tags around images other people have posted of you.

Facebook users tagging friends without consent moves the muddled state of online commerce to a whole new level. Now the Stockholm syndrome of people identifying with their captors has shifted, so consumers are inadvertently expediting the rise of a personal data economy dictated by advertising. Our complacency toward behavioral targeting means we're not only giving our data away for free but also accelerating and improving how advertisers and strangers access the digital breadcrumbs of other people in our lives.

I interviewed Kaliya (aka Identity Woman), executive director of the Personal Data Ecosystem Consortium, on this idea about losing control and giving data away and she noted, "The metaphor of slavery or feudalism is appropriate—the power dynamic between 'us' and the institutions that have our data is the problem that needs to be rebalanced, and until it is—we are slaves to our digital masters."[7]

The Future Value of Your Connected Life

The idea of data as related to identity has shifted in the past few years. Most of us grasp the concept of our online actions being tracked by cookies, or the notion that GPS can track our location. But the rise of wearable devices like Fitbit or the Nike+ FuelBand has begun to show average consumers visualizations of their data like they've never seen it before.

For athletes or the health-conscious, workout tracking used to rely on things like stopwatches and clipboards. Now sensors in wearable devices output data that can be tracked passively. This means users don't need to constantly input the specifics of their activities—devices do this for them. Connected to an iPhone or directly accessible by their doctors, this real-time aggregate assessment data is creating a revolution in the health industry. But it's also creating a stir akin to the online situation I've described above: Who owns your data? Who can sell it? How is it being used?

The trend of tracking data via wearable devices is commonly referred to as the quantified self (QS) movement (a term coined by Kevin Kelly and Gary Wolf of *Wired*), which typically involves individuals knowingly measuring their own behavior. This passive tracking ability has also moved offline to the devices and world around us, a trend known as the Internet of Things (IOT). There are a number of variations of this term, like machine-to-machine (M2M) technologies being utilized by the auto industry, or the Internet of Everything as coined by Cisco. The combination of QS and IOT results in a vast scope of information being recorded about our lives at all times, a trend referred to as Big Data.

What's so critical to understand about this data evolution is the logical transference of existing online norms around privacy and advertising that will likely be utilized in the *Outernet*, the virtual extension of the Internet that exists around us at all times but remains hidden from sight. The advent of a technology called augmented reality (AR), however, changes this dynamic as it allows for people to see data overlaid on the screen of their mobile phones or lenses of a device like Google Glass.

While many tend to focus on the applications of augmented reality mainly for gaming, I see its primary significance as a browser for this impending virtual world. I've written about AR since 2009 and its many benefits for business, health, and commerce, but, like any technology, it's the context of how it's used and by whom that is critical to establishing trust.

Google has faced numerous instances of eroding trust over the past few years relating to data usage and privacy. In 2012, the Federal Trade Commission fined Google $22.5 million for bypassing privacy settings in the Safari browser, the largest civil penalty ever levied by the FTC. And in March 2013, Google publicly acknowledged that it violated people's privacy during its Street View mapping project, as reported by David Streitfeld of the *New York Times*.[8] The company agreed to settle the case with thirty-eight states that rejected the unauthorized collection and use of people's data.

The case is of special interest as a precedent with regard to Google's Glass technology. If Google had no problem outfitting cars to collect people's data, why would they fret over wearable computers that can do the same? And if users have become complacent over privacy, as demonstrated by Facebook Graph Search behavior, then all the better for companies like Google and Facebook. People can continue to be the free conduits of evolving online commerce, believing their data is worth the cost.

It's when augmented reality becomes ubiquitous that these issues of data privacy and commerce will become visible. They'll literally be right in front of our eyes. And a final technological component people are already utilizing will cause widespread cultural concern. Facial recognition technology lets the user point their phone or device at someone's face and instantly obtain their name and other available data. And remember—Facebook has allowed millions of people to tag themselves and others for years in pictures they've posted to their pages. While the idea of crowdsourcing users to stay in touch with and tag friends may or may not be of concern to you, what will likely be upsetting is when strangers can access photos and data instantly by simply looking at you in public. Remember the tagging idea from Facebook Graph Search?

Now strangers will see those pictures floating above your head while you're waiting in line at Starbucks.

The good news about this existing digital economy is that many consumers have stopped being complacent about the misuse of their data. A recent Edelman study[9] found that 90 percent of consumers are concerned about the data security and privacy of their online information, and roughly 70 percent reported that privacy and security was a concern they had regarding their social media accounts. In the same study, respondents stated that they were, on average, 67 percent likely to switch their social media providers or stop using such services entirely if their information were accessed without permission.

So there's a reason to Get H(app)y—people are beginning to feel their data has renewed value, and are claiming the right to know how it's being used. There are dozens of companies like Personal or Reputation.com providing a model of data for people to store all the elements of their digital identity safely in one place, a model known as "personal clouds" (also called "data banks" or "data vaults"). Likewise, there is a greater sense of desired accountability from companies who are, by and large, controlling the data economy.

But here's a sticky wicket—the same standards of accountability we apply to Facebook and Google will be measured against us. How we conduct ourselves in the realm of the Outernet will leave digital fingerprints that will define our character while leaving a trail of identity-defining data. I refer to this trend as the rise of accountability-based influence (ABI), in which scores similar to eBay's detailed seller ratings gauge individuals' actions versus their words. Tracking trust will soon become akin to seeing a person's credit score, and the lines between Klout and commerce will blur even further. Will people benefit more from being popular and having influence, or by demonstrating positive character as defined by the digital portraits of their actions?

Both models will likely evolve in unison. And the growing

adoption of virtual currency platforms will mean people will become more comfortable with the idea of exchanging specie (market-
based money) for currency (social capital in the form of trust or
influence).

The Science of Happiness

A maturing field of science known as positive psychology is helping people see themselves in a new light. Measuring ourselves by
our virtuous potential rather than focusing on our brokenness is
transforming the nature of therapy in the modern world. We all
have pain, but it doesn't have to be a stigma—actions and behavior
associated with that pain are also *data*. When allowed the opportunity to optimize our lives unhindered by condemnatory scrutiny,
we can use data and new digital tools to make ourselves happier.

This science of happiness, which encompasses the fields of
psychology, physiology, and economics, is proving that we aren't
born with set levels of well-being. Unlike the medieval idea of
humors, we're not predetermined to suffer throughout our lives
based on rudimentary assessments of temperament. Some call it
"well-being." Some called it "eudaimonia." Some call it "flourishing."
Some call it "flow." Some call it "life satisfaction." However you
phrase the idea of a deeper, intrinsic, and long-lasting increase of
happiness, I have great news—you can increase it no matter who you
are. Science is proving this fact. When we better protect and manage our personal data, we'll also be able to decide who gets access to
and benefits from the specific attributes of our emotional lives.

Happinomics—or the Economics of Happiness

There is an economy of happiness. This isn't figurative. In the
sense that our actions, moods, and collective behavior can be
tracked with greater nuance than ever before, monetary and policy

decisions can be made that affect the economic standing of a population. Note that "happiness indicators" in economics typically don't refer to just the mood of a country—these indicators are metrics referring to multiple aspects of "well-being" which typically comprise details about things like the environment, education, and physical and mental health. Countries such as Bhutan, the United Kingdom, and the United States have all begun exploring how these metrics, which measure a wider breadth of attributes than GDP (gross domestic product), can give a clearer picture regarding the health of their citizens. These indicators typically focus on measuring increased well-being, a term that goes beyond mood and refers to a state of balance between multiple factors that affect you overall. On an individual level, well-being comprises your physical, mental, and emotional health. On a national level, well-being examines issues like education, the environment, civic engagement, and citizen health.

Multiple experts who study the science of happiness believe the positive increase in mood many associate with happiness comes as a result of action. In a sense, happiness is an output you experience after achieving a goal. On a national level, metrics gauging happiness are utilized to best understand how the actions of a government are improving people's lives.

Does it seem strange to measure people's happiness as an indication of a country's success? It is a newer idea, but has become a global trend because of increased sentiment that the measurement of gross domestic product (GDP) isn't working. First developed in 1934 by Simon Kuznets, a Russian-American economist, the GDP metric was adopted as the main tool for measuring a country's economy in 1944 after the Bretton Woods conference. This gathering of 730 United Nations delegates was tasked with trying to regulate the global economy after the Second World War. And in the sense that the model was adopted globally and used as a standard measure, it's been helpful.

But the GDP is primarily focused on financial measures—things like increases in goods production and salary levels. It doesn't account for the quality of a country's educational resources or care of the environment, and it wasn't designed to. In many ways, the GDP has become the primary measure of a country's success, casting a value judgment on citizens based primarily on wealth.

But a focus on increased productivity as measured by the GDP hasn't increased happiness. As Jeffrey Sachs, the renowned economist from Columbia University and one of the editors of the *World Happiness Report* created for the United Nations, notes: "The U.S. has had a three-time increase of GDP per capita since 1960, but the happiness needle hasn't budged."[10] Happiness measurement in this regard is based on something called subjective well-being, which means asking people to rate their happiness on a numbered scale.

It's remarkable that the GDP has become such an influential measurement of value considering all the things it doesn't account for. Famed New Zealand politician and author Marilyn Waring pointed out in her book *If Women Counted* that the GDP systematically underreports work performed by women who take on the traditional role of primary caregiver in the home. In this context, in a very real sense, according to the GDP's assessment of value, *women don't exist.*

What we choose to measure matters.

I want to pause here and remind you of something. The measures of subjective well-being and GDP produce data. Traditionally these metrics have been collected largely through survey responses. But with the advent of social media, wearable devices, and ubiquitous computing, capturing happiness data will become commonplace. This is significant because in a world where fiscal wealth has been the predominant measure of value, the hidden strengths and attributes of people as revealed by technology will allow for a form of "merrytocracy" based on personal design. Quantified happiness, determined by individuals, will begin to drive a new form of economics based on data.

Let's be clear—in the same way that I can't tell you what makes you happy, no technology can necessarily fully quantify your emotional state. But technology can provide what's known as a "proxy" for behavior or emotion, and it's the insights gained from these examples that can imply happiness or well-being, especially as related to health.

As an example, the Georgia Tech Homelab did a study with seniors living at home alone in which a simple sensor was placed on their bathroom doors, indicating when it opened or closed. Over the course of a longitudinal study they discovered that a 1 percent increase or decrease in the movements of that door suggested upwards of a 50 to 60 percent deterioration in health. Bathroom visits are a prime indicator of physical well-being—fewer visits could indicate bloating or more frequent trips could suggest dehydration.

Mobilyze is a platform designed to help people suffering from depression, employing mobile technology and context sensing to help with time-specific interventions. As described in the *Journal of Medical Internet Research*, the team behind the project developed an app where algorithms predicted patients' moods, activities, and environmental and social contexts based on thirty-eight different types of mobile sensors.[11]

The study, conducted in 2011, was one of the earliest attempts to use context sensing to identify mental health–related states. The relevance of context here also means the insights that can be gained by comparing different sensor data. For instance, did a patient register a more negative mood indoors than outdoors? Noted once, this observation is merely information. If the behavior is repeated, a caregiver could suggest a patient spend more time outdoors to quantify whether that behavior would have ongoing positive results.

The algorithm reference for the app above is also critical in this description. If any quantified behavior is repeated on multiple individuals in a study with enough data to categorize a pattern, a machine-learning model (algorithm) can be generated. This algorithm could help identify behaviors in new users and predict how they may react—in a sense, the tool "gets to know" a patient and can help them *before they even have a negative incident*. Combined with a voice recognition and search platform like Apple's Siri, the "personal digital assistant" model is evolving rapidly. Where patients or individuals are part of shaping how their data is studied, privately and in context, this technology is transformative. When people are kept out of the loop, and intimate data is exploited, algorithms may be driven solely from the intention of profit versus benefit.

As a final example, a research team at the University of Cambridge built Emotion Sense for Android, an app that lets you "explore how your mood relates to the data your smartphone can invisibly capture as you carry it throughout the day."[12] Using the highly articulated microphone in an Android phone, the app identifies multiple emotional states from users based on the inflections of their voices.

This passive quality of information capture, where users don't need to actively input data, is what is transforming modern measurement of people and their emotions. A report written by the researchers, "Emotion Sense: A Mobile Phones–based Adaptive Platform for Experimental Social Psychology Research," emphasizes this point: "Mobile sensing technology has the potential to bring a new perspective to the design of social psychology experiments, both in terms of accuracy of the results of the study and from a practical point of view. Mobile phones are already part of the daily life of people, so their presence is likely to be 'forgotten' by users, leading to accurate observation of spontaneous behavior."[13]

Until now, anyone attempting to measure emotion via traditional means has always confronted the problem of what's known as survey bias—people respond differently to questions when they know their responses are being measured. Passive sensors and ubiquitous computing mean that the objective or quantified measure of people's data will become more accurate. And the subjective assessment of their well-being (asking if they're happy) will still be utilized to determine people's responses to whatever is being measured.

What I've demonstrated with these examples is that we're in an era when people are beginning to realize that technology will help them accurately assess their emotions and the actions that contribute to them. We've all thought we were in love with someone who turned out to be wrong for us. We've all had jobs we thought would be great and we ended up not being fulfilled by the work. What if you had a Mio Alpha watch that measures heart rate that you wore on multiple first dates to see who, literally, made your heart skip a beat? What if the watch worn at work could help you identify the stress patterns brought on by an abusive manager, so you'd know the real reason you didn't like your job?

The hidden is becoming visible. Culture will shift. The rules have not yet been set.

So let's recap—people are now tracking their own behavior, moods, and health via quantified self tools. Objects around us (our cars, appliances) are outputting data directly related to how we move through the world. Soon, devices like Google Glass will let consumers record the world around them, tagging and posting massive amounts of content further relating to people's data.

Are you beginning to see the future I envisioned during my epiphany with Klout? Like the GDP's singular focus on wealth creation to determine value, Internet economics will continue to be driven by the accelerated exploitation of consumer data if a newer model isn't adopted soon. Behavioral targeting via passive tracking means the intimate measure of all your actions can more easily be utilized for sale. The data economy is more personal than ever, and if things don't change, our identities will be determined by algorithms controlled by someone else.

The H(app)y Hypothesis

Socrates said that the unexamined life is not worth living. The bad news is, your examined life, in the form of data, is worth *selling*. So if you want a say in the future of your identity in the Connected World, complacency is not an option. It's time to take action.

Here are the three parts of *Hacking H(app)iness*:

- A—be Accountable
- P—be a Provider
- P—be Proactive

PART 1: BE ACCOUNTABLE (IDENTITY AND MEASUREMENT IN THE CONNECTED WORLD)

The first part of living an examined life in the digital world is to understand how you're represented within it. In Part 1 we'll discuss the nature of connected identity and compare the trend of social influence (Klout, or "word-based sentiment analysis") to accountability-based influence (digital representations of action and trust). We'll explore how the emerging field of personal identity management provides a way for consumers to protect their data while maintaining flexibility in how they want to project their digital identity.

Then we'll get our geek on and explore the role of sensors, quantified self, the Internet of Things, and artificial intelligence as they relate to identity and happiness. We'll spend some time discussing the effects of machine-learning algorithms and how they relate to our digital future, and conclude by reviewing how our actions reflected in the connected world reveal a clearer portrait of identity than our words alone.

PART 2: BE A PROVIDER (BROADCASTING VALUE IN THE PERSONAL DATA ECONOMY)

There's a relationship created when we think of ourselves as consumers—while the word reflects the fact that we live in a transactional society, is it the primary identity we want for ourselves? A primary way to escape exploitative practices (like our tracked behavior being used primarily to enhance advertising models) is to change the vocabulary around an established idea.

In this section, we'll discuss how the concepts of shared value and conscious capitalism relate to the connected world. Where people's data is seen as commerce, its value should be distributed. In the personal data economy that will be made visible by augmented reality, we can inspire innovation while honoring privacy.

Rather than worrying about strangers filming and tagging without permission, people can broadcast their identities in public while notifying how they'd like to interact with the world. If you're at Starbucks and someone looks at you wearing Google Glass, your digital avatar could appear in their vision and say, "If you'd like to record and I'm in your shot, my face will appear blurry and I can't be tagged without my permission. If you're tagging me for commercial purposes, please text me the specifics of how I'll be compensated for the use of my personal data."

This type of scenario, outside of the technical aspects, represents the rapidly emerging practice of virtual currency. Within a trusted framework, people can pay each other in the form of specie (money), products (swapping), or skills (time). This avoids the echo chamber of privacy discussions mired in policy in favor of positive economic exchange. This is also a vision of how we can shift the model of selling people's data without their knowledge. We can shift this practice from being exploitative to being inclusive by providing transparent means of identity sharing and virtual commerce. Then people can see themselves as providers of content or data, where they are actively involved in a consensual transaction. The notion of being a consumer, defined primarily by what and how much is purchased, will erode and allow people to see their value in a wider dimension.

Geekery in this section will involve the evolution and future of augmented reality, a definition of Big Data, and how providing content and value to others can liberate your identity through creativity and commerce.

PART 3: BE PROACTIVE (PROMOTING PERSONAL AND PUBLIC WELL-BEING)

Many times, happiness is an output of action versus a momentary mood. Social scientists make the distinction between short-term or "hedonic" happiness and eudaimonia—a Greek term associated with Aristotle, roughly translated as "well-being." A new outfit may produce a momentary increase in positive mood, but if you rely on retail therapy for happiness you may experience what's known as the "hedonic treadmill."

Altruism also has proven benefits toward the increase of happiness. As Sonja Lyubomirsky, a leading mind in the field of positive psychology, notes in her book *The How of Happiness*, one of the less-noted aspects of kindness is its benefits regarding self-perception. The more acts of compassion you perform, the more you view yourself as altruistic. Eventually, the way you view your own identity may evolve to the point at which your confidence and happiness increase as a result.

Sharing value in the connected world leads to happiness. Also, the alternative isn't great—if you're a jerk, your actions may get quantified in a way to let others know that before you even speak to them. I wrote about this potential culture clash in my Mashable piece "The Impending Social Consequences of Augmented Reality."

Private data revealed in a digital context via technologies like augmented reality is going to lead to a lot of awkward situations:

> Ford's MyKey technology, available since January 2011, lets parents program cars for teens so they can't go over 80 mph or listen to the stereo until all seat belts are engaged. While the features were originally designed for teen safety, the technical framework could certainly be utilized in a different context . . . for instance, to vet whether or not a parent is worthy of driving children in a car pool. If via my "You Drive Like an Asshat" app I see you score a two out of ten on safety, my kid doesn't get in your car.[14]

This example demonstrates how accountability-based influence could become a key driver of identity and behavior in the future. In one sense, we'll start labeling other people the same way we rate restaurants right now in a Yelp review. And if no ethical or cultural frameworks around privacy or etiquette exist, data taken out of context will become almost a daily occurrence. That's why, in this section, we'll also be discussing the idea of "regard," or why it's so important not only to put your device away when speaking to someone face-to-face but to study how our interactions are different in the real and virtual worlds. Both have their benefits, but research on the longitudinal results of Facebook and other social network usage are showing negative effects that can be minimized by unplugging the connected side of your identity once in a while.

We'll examine thinkers from the world of positive psychology, focusing on how action, or "flow," as Mihaly Csikszentmihalyi (pronounced "cheek-sent-me-hi-ee") describes it in his seminal book *Flow*, can produce "optimal experience" in a person's life. By identifying the activities that drive your intrinsic well-being, you can optimize and improve the quality of your happiness.

Last, we'll focus on the emerging economic metrics of happiness indicators as demonstrated by Bhutan's Gross National Happiness Index. Other countries around the world, including the United Kingdom, Brazil, and the United States, are beginning to implement subjective and quantitative elements of policy based on measuring well-being.

I'll point out multiple examples of how the GDP isn't working as a measure of happiness, such as Shirley S. Wang's article "Is Happiness Overrated?" where she cites a 2010 statistics report in *Clinical Psychology Review* by researchers at San Diego State University, who noted that depression and paranoia had increased in college students from 1938 to 2007 and comments, "The analysis pointed to increasing cultural emphasis in the U.S. on materialism and status, which emphasize hedonic happiness, and decreasing attention to community and meaning in life, as possible explanations."[15] The popular book *Quiet: The Power of Introverts in a World That Can't Stop Talking* by Susan Cain touches on similar trends, showing how America has moved from a culture of character to one of personality. Our need to demonstrate extrovert characteristics has made us into a nation of salespeople, focused on self-aggrandizement over the benefit of others.

I'll also show the connection between the digital metrics of quantified self and the Internet of Things and the economic measures of Gross National Happiness. In this way, people can better connect their personal actions with a new global paradigm of value that's not based solely on wealth. Sharing value, done proactively, can provide individual happiness while changing the world for good.

The measurement of life based solely on fiscal wealth, or ever-increasing production or consumption, limits who we are. We aren't just creatures put on the earth to amass stuff or work ourselves to death. The economic measure of gross domestic product has influenced our lives to the deepest level of our global identity. Sadly, the existing data economy reinforces the fundamental tenets of GDP's focus on increased productivity at all costs (pun intended). When our lives are measured primarily as a marketing algorithm, we stop valuing actions that don't add up to a fiscal bottom line. We can't give ourselves permission to deeply reflect on what brings our lives meaning, or put others first when they need help.

But here's some great news—this primary measure of value the world has agreed on for more than fifty years is beginning to crumble. While the GDP, on one hand, is simply a metric to gauge the health of a country, it has so influenced our collective lives that most of us gauge our work not by its value but by its volume. We're not encouraged to take the time to see all the areas of our lives that can bring ourselves and others joy. We're not leveraging our full resources as humans and suffering due to the deficit.

But around the world countries are beginning to measure their citizens' lives and governmental actions via a wider lens. Multiple factors beyond financial metrics are being evaluated to see how people can live balanced lives beyond solely monetary measures. And when people gain perspective on all the ways their lives bring value beyond money, they'll also justify taking time to optimize their own lives or help others. They'll be motivated to take actions to increase their well-being in ways they haven't considered since the invention of the GDP.

My Background in Measurement

Bullies made it easy for me, a fat kid growing up in a suburb of Boston, to begin a life of self-examination. Early on, I became part of a playground hierarchy that had a set of sacred measures. Being overweight meant you were bullied. Fat equaled bad. Pretty simple. I wasn't happy about the situation, but I couldn't control it. So I studied it.

I learned that words don't often mirror action or character. For instance, the bullies who threatened were typically the last ones to act. I also became intimately aware of the concept of morals—I felt it was wrong that I was bullied. It wasn't fair. Nobody asked my permission but I still got cast in a John Hughes movie where roles were defined by somebody else before I even entered the picture.

I bring this up in regard to my experience with Klout to demonstrate how often we find ourselves in situations where someone has developed rules for a game we didn't know we were playing. And a typical human response is to try and win at a game without even asking whether it makes sense. That's what I mean by the challenge of measurement. We tend to look at the world through the lens we're given, never asking how the glass is focused. Questions of comparison are only applied to an existing perception of the world, versus one that may not be seen.

I was exposed to other ideas about examining life at a young age because of my family. My dad was a psychiatrist in the 1970s, when the term "shrink" was still applied in a pejorative sense. People tend to forget psychoanalysis is a relatively new field, having gotten its start from Freud in the early 1900s. When I moved to Needham, Massachusetts, as a boy, no neighbors brought us pies until they saw my dad gently spank me on the butt and realized he was mortal.

While he wasn't allowed to talk about his work with me, I knew my dad's job was to listen to people and help them hurt less. His private practice ran for about forty years in which he spent at least fifty hours a week, fifty weeks a year, helping patients—and that's a low estimate. Do the math and that's over one hundred thousand hours helping others examine their lives to find happiness.

Heroes come in all sizes. I come from good stock.

I went to college thinking I was going to be a minister, having examined a number of spiritual issues and thinking I could best help the world from the pulpit. But an influential acting teacher told me to follow my bliss, and I ended up going to New York City in 1992. (As it turns out, my mom wound up becoming a pastor.) I was a professional actor for more than fifteen years, appearing in principal roles on Broadway, on TV, and in film. During my time as an actor, I did a lot of corporate videos and started getting asked to rewrite scripts to make them funnier. I said, "Yes, if you pay me," and my writing career was born.

After multiple scripts, articles, and books, I landed a job writing the first About.com Guide to Podcasting in 2005 before social media became mainstream. I interviewed hundreds of thought leaders in business and technology before shifting to consulting to leverage my expertise toward business development for a few start-ups, like Blog Talk Radio. I ran two open-source tech conferences in New York City (two thousand participants, over two hundred speakers) and eventually ran the social media practice of a top-ten global PR firm. It was at that time my business career got two giant boosts—I became a contributing writer for Mashable.com, and traditionally published my first book, *Tactical Transparency: How Leaders Can Leverage Social Media to Maximize Value and Build Their Brand* (Hoboken, NJ: Jossey-Bass/John Wiley, 2008).

My primary expertise is in technology, having helped a number of great clients over the years, including Gillette, HP, and Merck, interpret how their brands were perceived online to build relevant engagement with consumers. At Mashable, I've interviewed hundreds of thought leaders from companies like Microsoft, Qualcomm, Google, Starbucks, Weight Watchers, AT&T, Verizon, and dozens of start-ups.

So in many ways I get paid to observe. As an actor, you observe human behavior to portray characters as real people. As a nonfiction writer, you observe trends to see how they'll affect culture. In my case, it was examining the technology and economic trends I wrote about in my Mashable piece "The Value of a Happiness Economy," which led to the writing of this book and the formation of the H(app)athon Project, a nonprofit organization "Connecting Happiness to Action" by creating sensor-based smartphone surveys utilizing economic indicators to increase civic engagement and well-being.

Thank You

Your time is precious, the most valuable resource someone can offer. I genuinely appreciate your devoting some of it to reading *Hacking H(app)iness*. I've done my best to provide sound value in exchange.

I gave you a bit of my background to demonstrate why I believe it's essential to live an examined life, and how much of my career has been spent in teaching people how technology can improve theirs.

I believe in the inalienable right of dignity for your data. Data is the proxy for your identity, a mark of your citizenship and humanity visible to the world. It's what makes you count.

In this sense, *Hacking H(app)iness* is about life, liberty, and the transformation of pursuit. The *pursuit* is what leads to the happiness. Identifying what brings you meaning and purpose and connecting to others is how to increase your well-being. This process can be expedited through revolutionary new mobile and digital tools, but not without personal reflection. And not without becoming accountable about your data and identity in the digital world.

So I invite you to reflect. And note—I'm not here to tell you how to be happy.

I'm here to prove you're worth the effort.

Be Accountable

IDENTITY AND MEASUREMENT
IN THE CONNECTED WORLD

NEVER CONFUSE
MOVEMENT WITH ACTION.
—*Ernest Hemingway*

YOUR IDENTITY
IN THE CONNECTED WORLD

I have preserved my identity, put its credibility to the test, and defended my dignity. What good this will bring the world I don't know. But for me it is good.

VÁCLAV HAVEL

THE "CONNECTED WORLD" has three meanings in this book:

- The Internet
- The Outernet
- How we relate to one another as people

The Internet you know from using it on your computer. You're probably also used to it living on your phone, and the concept of your car having Internet access is becoming familiar. But as a rule many think the Web is something that turns on and off with our computers, as we've become so used to it being in our lives. Unless Wi-Fi is down, we turn to the Internet and it's there.

The Outernet refers to the combination of technologies affecting you away from your computer. Your mobile phone contains many of these—GPS, an accelerometer, a microphone. But there are also technologies referred to as the "Internet of Things," which include sensors and devices in cars, buildings, and the world around us that we typically don't see.

You're aware of how your behavior is tracked on the Internet. Cookies are placed when you visit websites, leaving a digital trail that reflects who you are. In the Outernet, tracking happens the same way, but you're less aware of its effects, many times because of its mundane nature. If you live in the Northeast, you likely have an E-ZPass, a device you place in your car to speed toll collection. Instead of slowing down to pay an attendant, you choose the lane that automatically deducts the highway fee from your bank account. You don't think, "I've just left a point of data reflecting my identity" when you do so, but that's the case. The time and location of when you were driving directly tied to your finances is registered in an instant. A little piece of you is recorded for the rest of time.

The more the world is connected around us, the more snapshots of our behavior are recorded to form a picture of our digital, or virtual, identities. And soon technologies like augmented reality, in the form of Google Glass or dozens of other wearable devices, will allow us to see other people's data. How this will happen culturally is less clear than how the technology will work, but suffice it to say your actions in the Connected World will start to more directly affect your reputation and your well-being.

Well-being in the Connected World has three components:

- Subjective well-being—how you perceive your happiness and actions
- Avataristic well-being—how you project your happiness and actions
- Quantified well-being—how devices record your actions reflecting your happiness

These are critical distinctions that relate to other central themes in the book.

Taking a subjective measure of your happiness means I ask you to rate your *perception* of your mood in the moment. Your response, indicated typically as a number on a scale,[1] reflects your truth. You may inflate your score due to survey bias, but when asked under the right circumstances, measures of subjective well-being provide a powerfully transparent way of assessing people's perception of the world around them.

Avataristic well-being refers to how we broadcast ourselves, largely in the realm of social media. As an avatar is a manifestation of your character or alter ego in digital format, you can nurture or ignore it as you see fit. But the more engaging content you offer to online followers, the higher search rankings you're likely to get. When people pass on what you say or share, your online influence increases as a reflection of activity associated with your identity.

This is the realm of Klout and dozens of other similar companies working to measure online influence. These tools provide the benefit of breaking through the massive amounts of online content to identify people focused on particular subjects who have a significant following. "Significant" refers to the number or scale of people following an influencer or the demographic makeup of those followers.

Measuring influence in this way is of particular interest for brands. Identifying potential evangelists for products and services can be as simple as approaching top influencers in a category. Critical advertising dollars can go to increasing the scale of a message via an influencer versus building up a unique online following for the brand. Sentiment analysis of people's responses to messaging utilizing social analytics tools lets brands rapidly analyze how their marketing messages are faring with the audiences of the influencers with whom they've partnered. And all of it, the whole process, is tagged and ranked in search engines.

Klout and the like are not simply popularity mechanisms. Ranking via influence has direct impact on commerce, as reflected by where you appear in search. You may be the deepest-thinking mommy blogger on the planet, but if you don't show up on the first page of a Google search, then you're invisible.

Quantity of content is also a key indicator of influence. While it's assumed the quality of someone's tweets or posts is of a caliber to gain a large audience, the nature of following online is highly subjective. In this sense, we're encouraged to always create fresh content over and above if we have anything relevant to say. Increased productivity fuels the insatiable drive for salable data in the current model of the Web.

Christopher Carfi is VP of platform products for Swipp, a social intelligence platform company focused on measuring sentiment. He gave a talk with Robert Moran of the Brunswick Group for the World Future Society called "Rateocracy: When Everyone and Everything Has a Rating"[2] that touches on these ideas of influence. Moran's definition of rateocracy focuses on the ease with which people can create digital versions of their thoughts or feelings for others to see in the future—a global sentiment stream that can give the pulse of the planet. However, as Carfi points out, these sentiment streams provide a different form of benefit than influence measures like Klout.

There are a number of tools that use opaque algorithms in order to try to determine the level of "influence" an individual has in a social network. Unfortunately, these types of systems are easily gamed, often relating "influence" to frequency of social network posting activity. In contrast, future rating systems aggregate the feedback of large numbers of individuals to create a comprehensive picture of what various constituencies think about the world around them. Instead of trying to predict influence, the future of rating systems is more about better decision making at various levels of granularity. For example, if one was in the market for a new car, you might not care about what the auto critics think about that car. Instead, you might want to know what your friends who owned that model thought of it, or what the sentiment on that model was from others who had a family with similar demographics to your own.[3]

This form of crowdsourced rating system is becoming more common in the form of Yelp or other networks people have come to rely on in our digital age. However, modern behavior has still led a large majority of people to utilize digital sites to express opinions in hopes of peer approval. And when our primary focus for networking becomes the need to increase influence, we invariably suffer by comparison. *LiveScience* magazine describes this dilemma in "Facebooking and Your Mental Health." Author Stephanie Pappas cites recent research, pointing out the ironic fact that having too many friends on Facebook can lead to depression when people compare themselves with others' achievements. While the site itself isn't harmful, it's leading many users to feel worse about their own lives after using it.[4]

Here's a deep connection in the measures of online influence and wealth: When either is pursued only as a means to itself, people lose a balanced vision of what brings well-being. Klout and the GDP aren't evil. They're simply inefficient measures of holistic value for a person or a nation. The drive for increased influence or productivity alone becomes exclusionary for those who don't have resources relevant for improvement.

Kids are a part of the growing group of people without adequate resources to deal with identity issues in our modern digital environment. Along these lines, I interviewed Howard Gardner, Hobbs Professor of Cognition and Education at the Harvard Graduate School of Education, best known for his theory of multiple intelligences and education reform. His book *The App Generation: How Today's Youth Navigate Identity, Intimacy, and Imagination in a Digital World* provides deep insight into how kids relate to software applications and how apps can increase creativity and a strong sense of identity when viewed beyond the often limiting ways they're designed to be used. I asked Gardner how apps were shaping identity for kids in regard to influence issues.

> Social media like Facebook encourage individuals to "package themselves" so that they look as good, as perfect as possible. Not only does that inhibit exploration of identity, but it also makes youths feel inadequate, because they cannot live up to the apparently perfect lives of their peers. Of course, there were "role models" in earlier times, including ones taken from TV, but they were far more remote; one could not interact with them, and they did not change several times a week![5]

The stress involving upkeep of one's avataristic well-being is affecting us at an early age. Kids aren't allowed the time needed to understand identity issues, let alone know how to fabricate a perfect persona online.

Quantified well-being in the Connected World refers to how data is collected about a person that can be uniformly measured. For instance, the iPhone app Cardiio measures your heart rate as indicated by a change of color in your face via reflected light not visible to the human eye. The data is aggregated to a dashboard allowing you to share it or compare it with others. One can question the accuracy of the tool, but if it matches up to other modes of collection you can log your heart rate as an objective, quantified measure.

What becomes fascinating regarding your well-being and digital identity is when you start to compare more than one quantified data stream. Where it would be possible to create a longitudinal study, or repeated observations over a long period of time about the same variables in an experiment, you could start to see patterns that could lead to insights when comparing relevant data points.

A simple experiment would be running along the same route every day for a year and taking your heart rate at the same locations along the way. If you utilized your GPS data to track your location, time, and date stamps, patterns would emerge demonstrating which parts of the track spiked your heart rate. You'd see that the point in time at which you ran also affected your results (after a meal, late in the day). If you incorporated weather data in the mix, you might see how atmospheric pressure affected your runs and health.

I conducted a highly nonscientific test to show how emotions, or quantified happiness, may someday end up in the mix of measuring well-being. I used the Cardiio app and took my resting heart rate (see below) and got a reading of seventy-eight. Then, without moving from my chair, I watched Kmart's Ship My Pants[6] commercial on YouTube. (It's really funny; it involves multiple people saying "I shipped my pants" so you think they're saying something else.) My heart rate after watching the video once was seventy. By the third viewing (I was guffawing at this point), my reading was fifty-two.

Does this prove I've quantified happiness in some way? Not at all, but it's a start. Keep in mind I only used one data point—my heart rate—to measure my reaction to the video. If I used an app like Emotion Sense, I would also get time-stamped data relating to my mood and the tone of my voice. If I utilized a service like WeMo with home automation technology and utilized their motion sensors, I could send a text to my wife if I slapped the table hard enough while laughing. We'll get to the point when we may feel machines know our emotions better than we do.

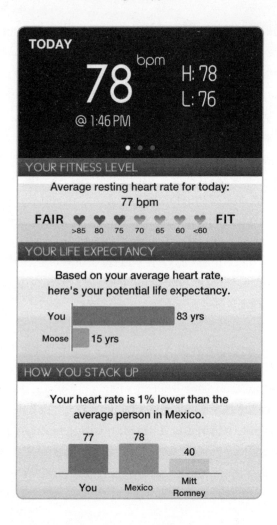

Note that my lowered heart rate finding due to smiling does have scientific precedent. In "De-Stress in Three Seconds," author Cassie Shortsleeve reports about a study conducted by Sarah Pressman and other researchers from the University of Kansas in which college students were asked to hold chopsticks in their mouths to simulate a smile before facing a stressful situation. Compared to their non-chopstick counterparts, smilers had lower heart rates and reduced stress, leading researchers to believe the physical triggering of facial muscles to smile sends a message to your brain that says, "You're happy—calm down."[7]

So what if I did a longitudinal study around my Cardiio experiment for a year, adding in whatever sensors I could think of? Would there come a point where I could prove to you that a certain amount of data proved I was experiencing a certain emotion? Probably. Keeping in mind, as in the case of taking a survey, that I know I'm recording myself and have a bias toward laughing.

But my point is not to quantify emotion for its own sake. My goal is to demonstrate how intimately connected we are to the data we're outputting and capturing in ways we've never done before. Mobile and home sensor technology is fairly cutting-edge, as is the trend of average consumers being able to capture their data. And remember, data is a currency. People pay for it. Think how Kmart's PR value would go up if they could prove that a thousand people doing my Cardiio test watching their commercial lowered their heart rates over time and significantly improved their health.

The idea may seem far-fetched until you hear about a company like Neumitra, featured in the *MIT Technology Review* article "Wrist Sensor Tells You How Stressed Out You Are," written by David Talbot. Neumitra has a device called bandu that's compatible with smartphones that can measure stress via increased perspiration or elevated skin temperature.[8] To research the piece, Talbot wore the device and tried to recite the alphabet backward in front of a group of strangers, resulting in his stress increasing by 50 percent as measured by wrist sweat and temperature.

Companies like Affectiva have also developed technology along these lines to recognize human emotions in the form of facial cues, letting brands test to see whether ads are engaging with consumers.[9] Much like my Kmart example, if a thousand people watching a certain video don't laugh as measured by Affectiva, it's a good sign the commercial is a clunker. I see this model moving to the social TV arena,[10] which is the trend of people interacting with live television programs or with other fans during prerecorded shows.

Whether facial recognition technology employed by Affectiva or Microsoft Kinect is reading our expressions during a show or our phones are measuring our reactions, our emotional output will be captured in one form or another. In terms of measuring stress, I think about watching a show like 24 and wonder at what point the TV would shut off if my heart rate got too high. Or when I'd get a call from my insurance carrier telling me to watch *Modern Family* to calm down before my rates got increased.

The emergence of quantified tracking of behavior signals that the avataristic form of well-being is fading in importance. While people will always follow influencers and repeat what they say, as we grow more comfortable with our actions being tracked, we'll be able to quantify emotions, or at least agree on the proxies for emotion based on physiology. Our actions will reveal our true characters. And reputation will more closely mirror our true selves versus the avatars we currently broadcast to the world.

ACCOUNTABILITY-BASED INFLUENCE

In the twentieth century, the invention of traditional credit transformed our consumer system and
in many ways controlled who had access to what.
In the twenty-first century, new trust networks and
the reputation capital they generate will reinvent
the way we think about wealth, markets, power,
and personal identity in ways we can't yet even
imagine.[1]

RACHEL BOTSMAN

MY MOM RECENTLY decided to move, now that it's been two years
since my dad died. After decades of meticulous financial record-
keeping and making payments on time, she learned she had to
restart her credit score from scratch as a widow. Reminiscent of
the gaping flaw in the GDP of not measuring women as primary
caregivers, this practice also highlights the need to overhaul an
outdated system.

Credit reporting's history began more than a century ago,
beginning with small retailers banding together to trade financial
information about their customers. The early credit associations
often focused on collecting negative financial information about
people as well as data about sexual orientations and other private
behavior. Oftentimes it was this private information that would
justify associations' denying services, reflecting negatively on people's reputations.

Not exactly a hallowed past regarding our financial forefathers.

Just as harrowing as this fiscal bigotry from credit associations was their lack of transparency. As Malgorzata Wozniacka and Snigdha Sen noted in their article "Credit Scores: What You Should Know About Your Own," it wasn't until 2001 that people could gain direct access to their credit scores.[2] This created a precedent for opaque collection practices around consumer information that data brokers have emulated in the online world. We have time in our Connected World, however, to wrest data back from brokers and control our identities and fates.

I wrote a piece for Mashable in 2011 called "Why Social Accountability Will Be the New Currency of the Web."[3] I was fascinated with online networks that had measurements reflecting trust generated by action where ratings were based on what people had done versus just how they were perceived as people.

One of the first places I looked was in the business world. Measuring performance is not a new idea, but typically it's only managers who rate employees based largely on their productivity. New models have emerged, however, that aggregate peer-to-peer comparisons to form a picture of someone's overall accountability, or reputation. One of these is Work.com, formerly known as Rypple and now a part of Salesforce.com. For my piece, I interviewed Nick Stein, who, at the time, was director of content and media for Rypple and is now senior director, marketing and communications, at Salesforce.com. A "social performance" platform, Work.com aggregates positive feedback (in the form of recognition) provided by colleagues. This recognition appears on an individual's social profile, providing a snapshot of that person's capabilities—as determined by their peers—thereby contributing to their reputation at work.

I asked Stein if he saw a day when someone's Work.com score could become portable, meaning it would follow an employee from one job to the next. While he felt the number of variables dependent on the context of one organization might not translate to a second one, he did feel measures like Work.com would have an influence on reputation.

> As we move toward a more social and transparent workplace environment, influence is becoming less dependent on your place in the org chart and more on the real, measurable impact you have on your colleagues. The idea is that all ongoing feedback, both positive and constructive, helps build an employee's real reputation at work . . . This enables individuals to develop influence based on their real impact rather than a perception of where they sit in the company hierarchy.[4]

I want to focus on Stein's idea of "real impact" now that the Connected World includes technology from sensors allowing quantified measurement. In the same way that brands will measure our emotional responses while watching TV, employers could track employees' moods or physical data as a reflection of corporate culture or performance. Right now it may be creepy to think of intimate data being visible to employers or peers, and employment policies need to protect information about sensitive medical conditions or other data people don't want to share in a work environment.

But let's examine the rise and implementation of social media in the workplace as a precedent for how sensor data could be adopted in the future. When social media first arrived, privacy was of huge concern but didn't keep the medium from becoming a mainstay of modern communication.

I began pitching the idea of blogs or podcasts for the business world in 2005, followed by Twitter, Facebook, and the like as soon as they became available. I sat in dozens of meetings where IT specialists warned about people using their own phones and devices at work, and managers expressed concern with employees wasting time on social media. I attended and spoke at hundreds of meetings and conferences discussing these issues and the merits of utilizing social media at work.

I think it's safe to say social media in various iterations has now been universally adopted for the workplace. Social media policies are in place, employees know the distinction between their public and work personas, and brands understand the vital importance of engaging with consumers where they get their media. I don't even use the term "social media" anymore—it's simply "media," and it's all social. The process of this adoption of social media in the enterprise has taken roughly a decade.

This precedent of technology focused on our identities in the form of social media will expedite the adoption of sensor data in the workplace.

In 2006, Hitachi approached the MIT Media Laboratory Human Dynamics Group to investigate the opportunities for "social sensors" in the enterprise. The resulting research in *Sensible Organization* provides a fascinating sense of how wearable and proximity sensors could affect the workplace. Part of the research focuses on how social sensors allow employees to visualize aspects of behavior revealed through these technologies. For instance, sensors could reveal which employees are more socially connected in an office, giving managers a way to quantify how best to disseminate communications to their organizations. Employees could also see how their activities could better dovetail with colleagues to increase their effectiveness at work.[5]

A more recent study of social sensors in the workplace from MIT, "Sensible Organizations: Technology and Methodology for Automatically Measuring Organizational Behavior," provides other pragmatic examples of technology utilized to improve dynamics among employees. Of particular interest is the "meeting mediator,"[6] which uses sociometric badges (lanyards worn by employees

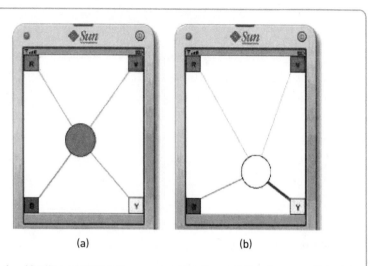

(a)   (b)

Fig. 1.   *Meeting mediator* is an example of a mobile phone–*sociometric* badge application. Each square in the corner represents a participant. The position of the center circle denotes speech participation balance, and the color of the circle denotes the group interactivity level. Fig. 1(a) shows a well-balanced and highly interactive group meeting, whereas Fig. 1(b) shows that participant *Y* is heavily dominating the conversation with a low level of turn taking between the participants.

measuring vocal tone and proximity) to help people understand how they're participating in a conversation. Displayed on a tablet or computer, employees are represented by shapes that equate to their physical location in a room. Depending on how often they speak and for how long, color patterns shift in the meeting mediator display, reflecting patterns in the conversation.

When people get used to the dynamics of seeing their conversations visualized in this way, this could become a powerful tool to combat the well-documented effect of meetings being dominated by certain personality types.

Accountability-based influence will begin to look a lot different in the coming years with the adoption of these types of methodologies. But along with privacy concerns, let's focus on the positive for a moment. Think about the meeting mediator being adopted where you work. How many conversations have you not contributed to because a domineering colleague wouldn't stop talking? Or if you're like me (passionate and verbose), how would your work benefit from letting others contribute more to a meeting? If a visual aid encouraging them to speak would help them feel more appreciated, they'd be more likely to find meaning in their jobs. Taken to scale, multiple employees with this type of tool may feel more loyal toward an organization. This could result in lower attrition rates, providing a measurable return on investment (ROI) for an organization.

Similarly, I wonder how a device like Neumitra's bandu stress bracelet could be utilized for the workplace. Imagine if a particularly nasty manager's team of employees all wore the bandu for three months. Combined with proximity and voice analysis sensors, data could show when and how the manager interacted with his employees. Data from voice sampling might indicate a majority of meetings involved raised or angry tones. Data from stress sensors might indicate a majority of employees had stress levels that directly correlated with these raised tones.

It's not rocket science—the data is simply reporting a visualization of the manager's negative management style. More important, however, is the quantified data providing a record that the manager is adversely affecting the health of his employees. In the near future, measures of ROI or quarterly reporting will need to take into account whether a manager is contributing to the positive or negative health of their employees. Ongoing high stress levels could increase a company's health insurance rates. In a world with sensors, manager accountability gets quantified.

Let's take the opposite situation. A manager utilizing empowering feedback for their team uses the same technology, potentially coupled with a Cardiio heart monitor for added data value. In direct contrast to the other example, insights gained from time stamp data could show an employee who exhibited high levels of stress had their anxiety lowered because of meeting with their manager. If they measured their resting heart rate before and after the meeting, data might show lower numbers for employees correlating to better health. In this scenario, our positive manager could be rewarded for improving morale, employee health, and saving the company money on their premiums.

I learned about the mobile technology agency Citizen after reading about them in the *Wired* article "What if Your Boss Tracked Your Sleep, Diet, and Exercise?"[7] The company has begun utilizing various sensor tools to measure employee health as a way to improve productivity. I interviewed Quinn Simpson, user experience director at Citizen, to discuss how he maintains a balance between privacy and innovation in his work. The team testing the implementation of sensors with Simpson are voluntarily providing data about their health, recognizing that it will take time for some colleagues to feel comfortable sharing various data. The company benefits from a strong corporate culture and a young demographic that is comfortable utilizing social media.

I asked Simpson about the idea of sensors with managers who might be overly negative (this is not a technology they're currently providing, but might in the future). His perspective was on the potential value of measuring performance via multiple data points as a great indicator of employee burnout. Pushing staff too hard on an extended basis, especially in a creative setting like Citizen, could lead to high turnover and loss of productivity. As Simpson noted:

> Because we keep track of what projects people are working on, we want to be able to look at data regarding their output and see when they risk burning out. Correlating relevant information points like this means you know when someone is not being as productive as they could be. So if I'm working too long on a project, both my manager and I want to know.[8]

What's encouraging about this example is how sensors and data promote unity among the staff. If a manager can quantify when a star creative is heading toward burnout, they can ask the employee to go home early or head to the gym. They will have data supporting their actions that better sustain their organization for the long term. Likewise, if identity or reputation models exist in their organization, they may earn more trust from employees for making a smarter choice for long-term gain versus short-term profits.

The New Reputation Economy

While I've focused on measuring accountability at work, it's easy to see how the technology explained in my earlier examples can live outside the enterprise. In a world with finite resources, we may soon enter a time when we'll see "sensible governments" utilizing technology that measures citizen behavior in an effort to improve their lives. While it's easy to consider this to be a Big Brother situation, where we'll be spied on at all times, let me describe a more supportive scenario that could come to pass in our Connected World.

There's a one-hundred-million-ton collection of plastic particles eddying about in the ocean known as the Great Pacific Garbage Patch.[9] While the particles are very small and the ocean has high powers of self-restoration, we'd still be well advised to increase our focus on recycling. In the same way that Work.com has a rating for employees to evaluate peers, what would happen if citizens began evaluating one another based on their recycling efforts? Or what if the sensor environment around citizens could contribute to a person's accountability rating as well?

Let's say you buy a bottle of water at the convenience store. Mobile payment technology charges your debit card but also indicates the bottle is made of plastic from its bar code and should be recycled. A time and date stamp with that information is sent to your town's local recycling facility. Let's say the facility has done a study and gathered enough data to know that your town's average length of time between buying a bottle of water and consuming it is one week. After seven days, if you haven't recycled your bottle, you might get a text reminding you to do so. If you're storing the water, you could indicate that in your response.

Citizens who recycled on a regular basis might receive a tax break at the end of the year because their efforts meant the town would receive money paid for bottles returned in bulk. But if you opted to chug your water and chuck the bottle in the parking lot, you might get a small fine for not recycling properly. If you left the bottle in a wildlife preserve as indicated via your GPS, you might get a larger fine. If your "recycling reputation" dropped low enough, retailers might be banned from selling you certain products.

It's not fun thinking negatively. But it's also not realistic to think our actions in the Connected World won't have negative consequences depending on the context. Where a community agrees on the types of things to measure and how privacy can be respected regarding data collection, more positive than negative results will occur.

This notion of communal benefit is reflected in a focus on sharing with the rising trend of collaborative consumption, a term introduced in 1978. Rather than drive individual consumerism, collaborative consumption encourages distribution of goods, skills, or money in peer-to-peer networks to preserve natural resources and lower individuals' costs for items they can share. Companies like Airbnb encourage home swapping, and Zilok enables rentals between individuals for everything from power tools to game consoles. Sharing and rating provide a robust platform for accountability and reputation models to emerge, buoyed by the advent of the Web.

As the *Economist* notes in "The Rise of the Sharing Economy,"

"Before the Internet, renting a surfboard, a power tool, or a parking space from someone else was feasible, but was usually more trouble than it was worth. Now websites such as Airbnb, Relay-Rides, and SnapGoods match up owners and renters; smartphones with GPS let people see where the nearest rentable car is parked; social networks provide a way to check up on people and build trust; and online payment systems handle the billing."[10]

If personal data can remain protected in these systems of open innovation, the sharing economy is a powerful move toward fulfillment in the Connected World and a positive example of how accountability-based influence can foster community versus self-focused gain.

3

PERSONAL IDENTITY MANAGEMENT

I'm excited about where technology will take us. My biggest goal is to make sure that our privacy laws keep up with our technology. I want to make sure that all of the benefits that we see from new technology don't come at the expense of our privacy and personal freedom.[1]

SENATOR AL FRANKEN

SENATOR AL FRANKEN is chairman of the Senate Subcommittee on Privacy, Technology, and the Law, a bipartisan part of the larger Senate Judiciary Committee. It's a complex job to encourage growth of technology while honoring the nuance of consumer privacy. From the technology side, it's easy to dwell on how privacy advocates may hinder innovation and growth. From the privacy side, a loss of trust from previous violations combined with a lack of understanding about technology slows adoption.

Both sides have merit and need to be heard. But the issues need some context:

- People's *right* to privacy is different from a person's *preference* about privacy.
- Just because a certain technology *can* be built doesn't mean it *should* be.

Let's unpack these ideas a bit.

Privacy is tough to both define and measure. Depending on the context, the activity that's fine for one person may not be condoned for another. For instance, as a rule, most adults don't have a problem with the idea that websites collecting information from children under the age of thirteen should comply with the Federal Trade Commission's Children's Online Privacy Protection Act (COPPA), which requires verifiable parental consent for PII, or Personally Identifiable Information to be collected about children. Gathering this PII for younger kids outside of the parameters of parental consent is typically seen as creepy or worse. COPPA also covers ideas of how cookies or other tracking mechanisms should or shouldn't be utilized to collect behavioral data on kids.

But manipulating online systems of age recognition can be easier than you think, especially when parents help kids under the age of thirteen get onto sites. As the *Huffington Post* reported in their article "Under 13 Year Olds on Facebook: Why Do 5 Million Kids Log In if Facebook Doesn't Want Them To?" a *Consumer Reports* study conducted in June 2012 revealed that "an estimated 5.6 million Facebook clients—about 3.5 percent of its U.S. users— are children who the company says are banned from the site."[2] Surprisingly, many of the kids creating accounts are also getting help from their parents, according to the study.

Here's where things get tricky: If a site can't collect PII data about a user, it's very hard to identify their age. And Facebook does regularly eliminate the younger users it identifies. The article also notes that Facebook could lose upwards of 3.5 percent of its U.S. market, however, if it were more vigilant at keeping kids off the site.

In light of this article, I'd like to restate my second issue from above with a little tweak:

- Just because a certain technology *hasn't* been built doesn't mean it *shouldn't* be.

Facebook has bigger priorities than creating technology that can accurately identify if a person is genuinely under the age of thirteen. That's not a question—if 5.6 million users might be under the age of thirteen and Facebook isn't actively creating a technology to ban them as mandated by the Federal Trade Commission, by definition their priorities are clear. The fact they'd stand to lose 3.5 percent of their U.S. market if kids were bumped from the site also speaks to their priorities.

Parents helping underage kids to game the system are acting on their personal preferences. The fact remains, however, that parents are breaking the spirit of COPPA when they help kids under thirteen get on Facebook and that the company stands to benefit when these kids join the site. And now those kids will start getting tracked earlier, with their data being utilized or sold in ways they don't realize.

The fact that Facebook *hasn't* built a technology to accurately identify if someone is under thirteen doesn't mean they *can't*. And as they've created the largest pool of photographic data in the world identifiable by facial recognition technology, I think they'd be able to block kids better if they wanted. Their facial recognition technology launched as opt-out only (versus having users take the extra step to opt in), implying they don't want users to be able to opt out because it messes with their ability to monetize.

It's this lack of clarity around privacy that is fostering distrust from users and helping to create the personal identity management industry.

The Context of Data

Data is like your health. You don't really appreciate the way that data is being handled until something bad happens to you.[3]

> —WILLIAM HOFFMAN, director of the World Economic
> Forum's Information and Communications Technologies
> Global Agenda

Unlocking the Value of Personal Data: From Collection to Usage,[4] written by the World Economic Forum (WEF) in collaboration with the Boston Consulting Group, provides an excellent overview of the evolution of modern data collection and practices. The report was generated as a result of a series of global workshops conducted by the WEF over the course of a number of months in 2012.

One of the biggest difficulties about data collection has to do with the context of what it will be used for. For instance, in measuring health data regarding a particular disease or condition, knowing personal information about individuals in a trial will lead to greater insights than by using anonymized data. So a practice of always separating people's identities from the results of their trials or other contexts can hinder innovation.

There is also a technical issue with anonymization that any data scientist will remind you of: As a rule, it's often impossible to achieve. In an effort to demonstrate the need for consumer privacy, famed Carnegie Mellon researcher Latanya Sweeney showed that 87 percent of all Americans could be uniquely identified using only three bits of information: their zip code, birth date, and gender.[5]

For these and other reasons, WEF's report calls for a shift from controlling data collection to focusing on data usage. A primary reason for the shift is the evolution of Big Data, which refers to the exponentially large sets of information that need to be aggregated and studied before even knowing if they contain potential for insights. As the WEF report notes:

> Often in the process of discovery, when combining data and looking for patterns and insights, possible applications are not always clear. Allowing data to be used for discovery more freely, but ensuring appropriate controls over the applications of that discovery to protect the individual, is one way of striking the balance between social and economic value creation and protection.[6]

The distinction between data collection and usage is of huge importance. If you're able to protect your data to the point where no one can access an iota of your identity without your permission, how someone wants to collect it becomes irrelevant. It's the equivalent of a robber wanting to steal your money from the bank: Without your providing the key, your currency stays where you want it.

Forrester Research's report *Personal Identity Management: Preparing for a World of Consumer-Managed Data*, by Fatemeh Khatibloo, reflects the growing trend of people wanting to own and control their data. Khatibloo points out that consumers are beginning to better understand how marketers are making money off their data, and they're keen to learn how their data is being collected and used.[7]

The term "personal identity management" also reflects the need for consumers to shift from a complacent to a proactive stance regarding their digital identities. One key tool for them to leverage this shift is the rise of "data vaults" or "lockers" or "personal clouds."

A difficulty in how information gets shared about you has to do with the context of who is asking for your data and how long they need it to accomplish a mutually established goal. If you were able to collect in one place all of your personal information and data you generate as you use your devices, and control how and when it gets shared and under what terms and conditions, however, you'd be employing the mentality of a data vault.

Another key element with a vault is the idea of destroying data after a certain time limit or when it's used outside of the context the person sharing it intended it to be used. Reminiscent of the *Mission: Impossible* encoded spy message that self-destructs after being read, this idea of temporary data usage has caught on recently through the photo-sharing app Snapchat, where users allow a set time limit for pictures to be viewed by recipients. While the app has suffered from some users learning how to store photos longer than users intended, the trend of data destruction catching on could be a very positive one for consumers overall. People will begin to understand that a great way to protect their data is to only provide it to trusted parties and to enable it to be destroyed if they feel it's being used in ways they didn't permit.

Nobody's Vault but Your Own

"Some of the most open people, people who say they're open, change their minds when they learn how much of their data is beyond their control," said Shane Green, CEO of Personal, a company focused on helping users "take control of the master copy of their data" with services focused on protection and flexibility regarding digital identity about privacy in the digital age. "People are being digitized."[8]

Green is focusing on using a carrot-and-stick method to get users used to the idea of data vaults. One of the company's most popular offerings is a service called Fill It, which lets users auto-populate sign-in forms securely online. This provides a sense of how vaults work overall, as users see they're in control of their data. As Green noted in his interview for this book:

> People feel more protected when their data is protected. When you fill out a form, you're "turning over the goods." You're signing away your privacy and terms of use. This is the pain point everyone has. When you solve that problem, people see why they need a set of reusable data.[9]

The company also features a unique Owner Data Agreement on their site that turns personal users into owners. The agreement is a contract making users the legal owners of the data they store with the service. Overall the site provides a powerfully motivating message that reinforces the need for consumers to understand how precious their data is and take charge of it in a proactive way.

"I went to Harvard Law School and I can't understand most terms of service agreements," noted Michael Fertik in an interview for *Hacking H(app)iness*. Fertik is CEO of Reputation.com, a leading provider of online reputation products and services. In reference to data brokers or other sites not wanting to make it easy for consumers to understand byzantine terms of service agreements, Fertik says, "As the saying goes, if you're not paying for the product, you or your data are the product." While Fertik doesn't feel companies selling data are necessarily mendacious in nature, he points out, "They simply don't care about your privacy. Between the right thing and mammon, mammon will win."[10]

Fertik sees a future of protected consumer privacy via data vaults or similar services as being inevitable. Besides being ethically sound, there are powerful economic motivators for companies as well as individuals to innovate and evolve a personal data economy. A primary aspect of this motivation will come in flexibility of how companies will obtain consumer data in the future. Fertik outlines four primary types of exchange he envisions, where companies will provide the following to consumers in exchange for the right to access their data:

1) Coupons or discounts
2) Special privileges based on status (While this could come in the form of airline points, an evolution of this idea would be for influence or reputation points to be portable between airlines.)
3) Cash, virtual or real ("virtual" meaning virtual currency)
4) Privacy (People can make their purchases without revealing any data for a fee.)

These four examples provide a pragmatic approach to monetization in the personal identity management economy. While the model may cause anxiety for some brands reliant on purchasing customer information, the real people set to suffer in this system will be data brokers disintermediated in the process. For brands, this will provide more direct access to their customers and more opportunity to establish value-added relationships.

If consumers begin trading or selling their data in this type of model, brands also won't have to work as hard to advertise. Consumers will become masters of their own data and be able to meaningfully engage with brands that they want to interact with to create opportunities for mutual value creation.

John Clippinger, cofounder and executive director of ID3, a nonprofit headquartered in Boston whose mission (according to their site) is to "develop a new social ecosystem of trusted, self-healing digital institutions," has created the Open Mustard Seed (OMS) Framework as an open-data platform that helps users take and keep control over their personal information. Along with ensuring trust between users and anyone trying to access their data, Clippinger feels having user data in this form of "pool economy" means you can create new markets. As he explains in *"Power-Curve Society: The Future of Innovation, Opportunity and Social Equity in the Emerging Networked Economy,"*[11] reverse auctions (where users provide their data for sale instead of others using it without their knowledge or consent) could create "enormous efficiencies" in the future. Think about a model like Craigslist, where people list items they want to sell, specifying their own prices and specifics around the interactions. These data pools could provide that protected infrastructure.

The Personal Data Ecosystem Consortium, where Kaliya (aka Identity Woman) is executive director, is also creating an infrastructure that is helping to change the data market and make it real. Almost one hundred companies around the world are working on tools to help people collect, manage, and get value from their own data and build new ethical data markets. By joining the consortium, companies have to make a commitment to give people the rights to their own data and work toward interoperability so people will not be trapped by a particular provider.

If the personal data economy becomes widespread, consumers will also manage access rights to their data directly with one another. In a world of conscious consumerism, this might come in the form of an eBay power seller wanting to gain access to data about prospective clients from them. Or someone trying to establish credibility for renting their apartment on Airbnb may buy data from popular renters who are happy to monetize their expertise.

People will also start to earn passive revenue from watching TV while allowing advertisers to monitor their responses to shows using sensors or other technology. A precedent for this model can be seen from existing online video streaming sites. If you want to watch a thirty-minute video, some sites will offer the opportunity to watch a longer pre-roll ad to see the whole show uninterrupted. Or you can watch a number of shorter ads and have some breaks in viewing where you can't fast-forward over an ad. But the point is, you're given a choice.

Transparency combined with options in which people feel their data is protected means commerce can flourish because there's no need to hide shady business practices.

The Connected Choice

I think we see privacy violations where there are genuine gaps between what companies think is acceptable and what consumers expect in terms of privacy. For example, most consumers think that when they give an app permission to collect their location information, it's only that app that will get that data.

I think consumers would be shocked if they learned that half of top apps turn around and give or sell that data to third parties. Yet this is a standard business practice on the Internet—and many companies are sincerely surprised when privacy advocates raise it as an issue. So a lot of my work is simply shedding light on these practices and trying to bridge these two worldviews.[12]

—SENATOR AL FRANKEN

Hacking H(app)iness means breaking down old ideas of what can bring contentment in the Connected World. Fulfillment will only come when you recognize that your data is your own. It's an extension of your identity. If other people are collecting it, you should know what they want to do with it and be a part of the transaction if you so choose.

4

MOBILE SENSORS

The idea that the smartphone is a mobile sensor platform is absolutely central to my thinking about the future. And it should be central to everyone's thinking, in my opinion, because the way that we learn to use the sensors in our phones and other devices is going to be one of the areas where breakthroughs will happen.[1]

TIM O'REILLY

WHAT MYSTERIES are revealed by your behavior that you can't see? What would you learn about yourself if you could see the data reflecting your words and actions? If all your senses left a trail, if your actions painted a picture in data, how might that feel?

It would be magic.

Seeing your life visualized in data can be extremely empowering. Instead of fretting about the unfinished items on your to-do list, seeing your data lets you revel in the experience of your "I am" list. There's a weight to personal data, a permanence you can point to. The lines, curves, or numbers on a page, revealed in the form of data—that's *you*. Want to see if you're having an effect on the world? Visualize your output. It's a powerfully rewarding experience.

Sensors are the tools that interpret your data. Sometimes they're simple, like a pedometer, which counts your steps. Sometimes they're complex, like an MRI machine, which measures brain patterns. But they both provide insights that prompt action.

The first sensors we're made aware of in life are our bodies. They're highly articulated instruments, as they can feel, interpret, and respond to stimuli almost simultaneously. As children, we're exposed to sensors in a doctor's office, having our blood pressure taken and wondering why the scratchy black fabric of the monitor gets so tight we can feel our hearts beating in our arms. We get older and go through security and experience metal detectors. The concept of technology measuring the invisible is something we accept at a fairly young age.

But where we get skittish is when sensors begin to track us. We don't mind if air quality is measured for negative emissions, or if thermal patterns are tracked to portend weather patterns. But when things get personal, we flinch at first. It's a bit like scraping your knee and seeing blood for the first time; it feels like something that's supposed to stay inside of you has come out. Data has an unexpected intimacy upon revelation. It also cries out for comparison and action. You measure your resting heart rate knowing you'll gauge the resulting number against an elevated reading during exercise.

And what about Hacking H(app)iness? Can we use technology to identify and predict emotion? Would we want to if we could?

Affective computing is a multidisciplinary field deriving from the eponymous paper[2] written by Rosalind W. Picard, director of Affective Computing Research at the MIT Media Lab. Picard's work has defined the idea of measuring physical response to quantify emotion. While it's easy to focus on the creepy factor of sensors or machines trying to measure our emotions, it's helpful to see some applications of how this type of technology can and is already improving people's lives.

In the *New York Times* article "But How Do You Really Feel? Someday the Computer May Know," Karen Weintraub describes a prototype technology focused on autism created by Picard and a colleague that helped people with Asperger's syndrome better deal with conversation in social settings. The technology featured a pair of glasses outfitted with a tiny traffic light that flashed yellow and red, alerting the wearer to visual cues they couldn't recognized due to Asperger's (things like yawning that indicate the person you're speaking to is not interested in what you have to say).[3]

It's easy to imagine this type of technology being created for Google Glass. The famous American psychologist Paul Ekman classified six emotions that are universally expressed by humans around the globe: anger, surprise, disgust, happiness, fear, and sadness. Measuring these cues via facial recognition technology could become commonplace within a decade. Cross-referencing GPS data with measurement of these emotions could be highly illuminating—what physical location has the biggest digital footprint of fear? Should more police be made available in that area?

Picard's boredom-based technology would also certainly be useful in the workplace. Forget sensitivity workshops; get people trained in using this type of platform, where when a colleague looks away while you're speaking you get a big text message on the inside of your glasses that reads, "Move on, sport." Acting on these cues would also increase your reputation, with time stamps noting when you helped someone increase their productivity by getting back to work versus waxing rhapsodic about the latest episode of *Downton Abbey*.

"The Aztec Project: Providing Assistive Technology for People with Dementia and Their Carers in Croydon" is a report from 2006 documenting sensor-based health solutions for dementia and Alzheimer's patients in South London. The report starts off with the harrowing statistic that "there are currently some twenty-four million [Alzheimer's disease] sufferers, a number that will double every twenty years until 2040."[4] The scale of the population including patient families greatly increases this number, and the financial burden for all parties involved places significant stress on health costs.

My grandmother had Alzheimer's, so I identify with the scenarios described in the report. Wandering is a standard behavior, with patients not recognizing they've left or entered a room or even their home. Accidents in the kitchen are common, as is forgetting to eat for days on end. The report identified that previous solutions, including things like locking doors to keep patients from wandering or having them wear bulky lanyards outfitted with alarms, were ineffective. When lucid, patients felt trapped or tagged and resented feeling so scrutinized in their own homes.

Technical fixes due to advances in technology, even in 2006, provided solutions that brought great comfort to patients, their families, and caregivers. For instance, instead of having a patient wear an arm or leg band outfitted with a sensor detecting when they went beyond the radius of their home or property (a practice associated with criminals and upsetting to patients), an early form of geo-fencing technology was utilized instead that sent a warning text to caregivers when patients crossed over a virtual perimeter on their property. Sensors were also placed on doors that acted as simple alarms when patients left their houses unattended.

A more recent implementation of sensors to help treat Alzheimer's patients is taking place in Greece via a technology in development called Symbiosis.[5] Pioneered by a team from the Aristotle University of Thessaloniki's Department of Electrical and Computer Engineering, Symbiosis has a number of components to help patients and their families and caregivers. SymbioEyes incorporates the automatic taking of photographs via a mobile app also outfitted with GPS tracking and emergency detection capabilities. Worn by patients as a way to monitor location, the pictures are also viewed at the end of the day as a way to inspire memory retention and curb the onset of dementia. SymbioSpace utilizes augmented reality to create digital content that reminds patients of simple behaviors. For instance, the image of a plate triggers a text reminding patients how to eat with a spoon. Along with the pragmatic benefits of these reminders, they are designed to make a patient feel they are "surrounded by a helpful environment that provides feedback and seems to interact with him/her, responding to his/ her needs for continuous reminding and memory refreshing."

Kat Houghton is cofounder and research director for Ilumivu, a "robust, patient-centered software platform designed to capture rich, multimodal behavioral data streams through user engagement" (according to their website). I asked Kat her definition of affective computing and why sensors are so central to her work:

> Affective computing is the attempt to use systems and devices (including sensors) to identify, quantify, monitor, and possibly simulate states of human affect. It is another way in which we humans are attempting to understand our own emotional experiences. Sensors, both wearable and embedded in the environment, combined with ubiquitous wireless computing devices (e.g., smartphones), offer us a large data set on human behavior that has never before been possible to access.[6]

A great deal of Kat's work is focused on autism, where wearable sensors are being utilized for preemptive or just-in-time intervention delivery. "Using data from sensors to generate algorithms that allow us to accurately predict a person's behavior could radically change our ability to facilitate behavior change much more quickly and effectively," she says. Sensors can also play a role in identifying and preemptively intervening in states of what is known as "dis-regulation":

> When a person with autism (actually all of us to some degree) is in a state of physiological dis-regulation, they are more

likely to engage in challenging behaviors (tantrums, self-injury, aggression, property destruction, etc.), which cause a lot of stress to themselves and their caregivers and significantly restrict the kinds of learning opportunities available to that individual. We are using wearable sensors that monitor autonomic (involuntary) arousal levels in combination with momentary assessments from caregivers and data from sensors in the immediate environment to see if we can identify triggers of dis-regulation. If we can do this, then we are in a position to be able to experiment with providing preemptive interventions to help people with autism maintain a regulated state by providing input before they become dis-regulated. Right now the only option to caregivers is to try to offer support after the fact.[7]

As Kat notes, being able to know ahead of time what is going on for a person with autism could be a significant game changer for many of the more challenged people on the autism spectrum, along with their loved ones and caregivers. Sensors are providing a unique portrait of behavior invisible before these new technologies existed.

To find out more about the idea of interventions involving sensors and the tracking of emotions, I interviewed my friend Mary Czerwinski, research manager of the Visualization and Interaction (VIBE) Research Group at Microsoft.

Can you please describe your most recent work?
For the last three years we have been exploring the feasibility of emotion detection for both reflection and for real-time intervention. In addition, we have been exploring whether or not we can devise policies around the appropriateness and cadence of real-time interventions, depending on personality type and context. Interventions we are exploring include those inspired by cognitive

behavioral therapy, positive psychology, and such practices, but also from observations of what people naturally do on the Web for enjoyment anyway.

How would you define "emotion tracking"?
Emotion tracking involves detecting a user's mood through technologies like wearable sensors, computer cameras, or audio analysis. We can determine a user's mood state, after collecting some ground truth through self-reports, by analyzing their electrodermal activity (EDA), heart-rate variability (HRV), [and] activity levels, or from analyzing facial and speech gestures. Machine-learning algorithms are used to categorize the signals into probable mood states.

How has emotion tracking evolved in the past few years, in general and in your work?
Because of the recent advent of inexpensive (relatively speaking) wearable sensors, we have gotten much better at detecting mood accurately and in real time. Also, the affective computing community has come up with very sophisticated algorithms for detecting key features, like smiling or stress in the voice, through audio and video signal analysis.

Do you think emotion tracking will have its own "singularity"? Meaning, will emotion tracking become so articulated, advanced, and nuanced that technology could get to know us better than we know ourselves?
That's a very interesting question and one I've been thinking about. What is certainly clear is that most of us don't think about our own emotional states that often, and perhaps aren't as clued into our own stress or anxiety levels as we sometimes should be. The mere mention of a machine being able to sense one's mood pretty accurately makes some people very uncomfortable. That is

why we focus so much on the hard human-computer research questions around what the technology should be used for that is useful and appropriate, given the context one is in.[8]

Along with trying to demonstrate how technology is helping map and quantify our emotions, my bigger goal is in providing you permission for reflection. As Mary points out, many of us don't think about our emotional states, which means it's harder to change or improve. Note there's a huge difference between observing emotions and experiencing them, by the way. Observation implies objectivity, whereas in the moment it's pretty hard to note, "Gosh, I'm in a blind rage right now." So while it may take us some time to get used to tracking our emotions and understanding how sensors reveal what we're feeling, a bigger adjustment needs to happen in our lives outside of technology for the greatest impact to take place.[8]

Sensor-tivity

"We're really in the connection business." Iggy Fanlo is cofounder and CEO of Lively, a platform that provides seniors living at home with a way to seamlessly monitor their health through sensors that track health-related behavior. "Globalization has torn families apart who now live hundreds of miles from one another," he noted in an interview for *Hacking H(app)iness*. "Our goal is to connect generations affected by this trend."[9]

After investing in years of study, Fanlo came to realize that older people care more about *why* they're getting out of bed than *how*. So he focused on finding the emotional connections that would empower seniors while bringing a sense of peace to the "sandwich parents" (people with kids who are also dealing with elderly parents) concerned about their parents' health. Sadly, a primary reason for friction in these child-to-parent caregiver situations is that kids tend to badger parents to make sure they're taking their medications or following other normal daily routines. The nagging drives the parents to resent their kids and potentially avoid calling, even if they have a health-related incident that needs attention.

The company's tech is surprisingly simple to use, although its Internet of Things sensor interior is state-of-the-art. The system contains a hub, a white orb-shaped device the size of a small toaster. It plugs into the wall and is cellular, as many seniors don't have Wi-Fi or don't know how to reboot a router if it goes down. A series of sensors, each about two inches in diameter, has adhesive backing to get stuck in strategic locations around the house:

- Pillboxes—sensors have accelerometers that know when the pills are picked up, serving as proxy behavior of assuming parents have taken their meds.
- Refrigerator door—a sensor knows when the door is opened and closed, serving as proxy behavior for parents eating regular meals.
- Silverware drawer—a sensor knows when the drawer opens and closes, serving as a secondary proxy measuring number of meals eaten.
- Back of the phone receiver—a sensor knows if the phone hasn't been lifted, and after a few days, a message is sent to the child of the elderly person living at home reminding them to give their parent a call.
- Key fob—this features geo-fencing technology and indicates if a parent has left the house in the past few days.

A final component to Lively is that friends and family members contribute to a physical book that is mailed to seniors living at home twice a week. It's a literal facebook that adds to the emotional connection between generations.

The most powerful component of the platform, however, is the improvement of relationships between parents and their adult children due to a lack of constant nagging. "This was an unintended consequence we learned during testing," said Fanlo. "The seniors we asked were now happy to talk to their children. In the past, relationships had gotten toxic. The nagging was poisoning relationships. Since the children of seniors knew parent health was monitored, this took away the toxicity between the generations."[10]

Along with Big Data, the trend or notion of Little Data has been growing in prominence. Outside of the technical aspects of sensors and tracking technology being inexpensive enough for the general public to take advantage of, Little Data also refers to the types of interactions involving platforms like Lively. Data is centered around one primary node (the seniors) and their actions (four or five activity streams). It's intimate, contained, and highly effective at achieving a set goal to the benefit of multiple stakeholders.

Here's another aspect of sensors to note in these examples: They're invisible. Tracking doesn't always have to be nefarious in nature. Lively calls their health monitoring "activity sharing." For the next few years, we'll be aware of sensors in the form of wearable devices because they're new, much like we first felt about mobile phones when they were introduced. But after we become used to them, they'll fade from prominence and do their passive collection while we go on with our lives.

Disaster Data

Patrick Meier is an internationally recognized thought leader on the application of new technologies for crisis early warning, humanitarian response, and resilience. He regularly updates his iRevolution blog, focusing on issues ranging from Big Data and cloud computing to crisis mapping and humanitarian-focused technology.

Recently, he blogged about the creation of an app that could be

utilized during crisis situations to immediately connect people in need to those who could provide assistance. During crises, it can take many hours or even days for outside assistance to come to the aid of a devastated community. Meier is working to create solutions that can empower communities to provide help to one another in the critical time frame occurring directly after a negative event.

In his post "MatchApp: Next Generation Disaster Response App?"[11] Meier lays out the vision for an app using a combination of sensors that could help people both ask for and provide assistance during a crisis. (Note the image here is a mock-up; Meier also recently wrote[12] about an existing app called Jointly that has created a similar framework.) The concept of the MatchApp idea is quite simple: Like a shifting jigsaw puzzle, people's needs shift dramatically in real time in the wake of a crisis. But location, identified via GPS, plays a key role in helping match need with re-

sources in the most streamlined way possible. The figure on the previous page shows how a specific need is being met via a combination
of GPS (on the left) and a confirming text message (on the right).

Meier describes how privacy is maintained in this framework
while also providing a vehicle for increasing digital trust:

> Once a match is made, the two individuals in question re-
> ceive an automated alert notifying them about the match. By
> default, both users' identities and exact locations are kept con-
> fidential while they initiate contact via the app's instant
> messaging (IM) feature. Each user can decide to reveal their
> identity/location at any time. The IM feature thus enables
> users to confirm that the match is indeed correct and/or still
> current. It is then up to the user *requesting help* to share her
> or his location if they feel comfortable doing so. Once the
> match has been responded to, the user who received help is
> invited to rate the individual who offered help.[13]

The app and scenario provide a compelling example of how
implementing sensors can help improve our health and even save
lives. By proactively implementing protected data plans as Meier
has in the MatchApp, we also work around privacy concerns, as
people's preferences are taken into consideration and they're provided the choice to reveal their information as they see fit.

Margaret Morris—Left to Our Own Devices

Margaret Morris is a clinical psychologist and senior researcher at
Intel. She examines how people relate to technology, and creates
mobile and social applications to invite self-awareness and change.
In her TED Talk "The New Sharing of Emotions" (April 2013), she
discusses her work of creating a "mood phone" with her research
team at Intel. Designed to be a "psychoanalyst in your pocket," her

tool lets people self-track moods and other behavior in experiments when they are "left to their own devices." Morris's logic is that, as our mobile phones are always at our side, we can leverage them for insights to improve our well-being. The bond that patients form with traditional therapy can extend to our phones for connection to social networks or other resources.

Here's my favorite quote from her TED Talk: "We're at this moment where we can enable all kinds of sharing by bringing together very intimate technologies like sensors with massive ones like the cloud. As we do this, we'll witness new kinds of breaktough moments, and bring our thinking and all our approaches about emotional well-being into the twenty-first century."[14]

I interviewed Morris to ask about the issues she brought up in her talk, to see how people relate to their smartphones, where the technological aspects are less important than the feeling of being helped by having an omnipresent, trusted tool at their side. A number of these issues are also elucidated in an excellent paper Morris wrote with a number of other researchers from Intel, Oregon Health and Sciences University, and Columbia University called *Mobile Therapy: Case Study Evaluations of a Cell Phone Application for Emotional Self-Awareness.*[15]

Can you describe your work with the "mood phone"? How did that work come about, and how has it evolved?
I created the mood phone to show how tools for emotional well-being (e.g., those from psychotherapy and mindfulness practices) could move into daily life, be available to everyone who has a phone, and be contextually relevant. It started as a complex system involving wireless sensing of ECG, calendar integration, and just-in-time prompting based on cognitive therapy, yoga, and mindfulness. It emerged because I was asked (within my research group) to develop a new approach to technologies for cardiovascular disease. It was important to me to take a preventive approach,

focusing on psychosocial risks, and make something that would be very desirable and improve quality of life immediately while lowering long-term risk. I was interested in emotional well-being and relationship enhancement as motivational hooks for self-care. They are more palpable than long-term cardiovascular risk, and of course, the quality of our relationships affects everything.

Where do you see the balance between human psychoanalysts and the ones "in your pocket"? How can people determine that balance?

Most people do not have access to terribly good mental health care of any sort, much less psychoanalysis. They are "left to their own devices" and are remarkably resourceful, learning from friends, strangers, and using everything at their disposal, including their devices, apps, social media, and their own data.[16]

I think this last point is really important when considering why technology can be utilized to improve our well-being, whether it's mentally, physically, or emotionally focused. While it's understandable people would be concerned about replacing a human therapist with technology, it also doesn't make sense to ignore a tool we all have with us all the time that could help us examine and improve our well-being. Our mobile phones also provide us direct, real-time contact with our loved ones or people responsible for our care. Whether they provide aid in emergency situations or simply a reminder that people in our lives are looking out for us, it's also relevant to ask why we feel comfortable allowing these tools to track our behavior for marketing purposes, but get leery of using them to measure our emotions directly. Advertisers have no compunction about analyzing every decision, interaction, word, and action you take to get insights about the perfect timing to introduce their products or services. Why not utilize these same methodologies to understand your emotions in a personal context?

A final word on this idea that Morris mentions in her TED Talk. When her first experiment was done with the mood phone, she thought a lot of participants might have concerns about their privacy—how their data was being used, and so on. But what she heard most often was people asking if they could get the technology for their spouses. The insights the tools generated about their emotional lives led many of them to believe that their relationships would be greatly improved if they could take those insights and share them with the ones they love most. Elements of their personalities revealed by the technology created opportunities for discussions infused with an objectivity that wasn't available before the mood phone was put to use.

So my final question here would be, would you rather be "left to your own device" or continue to do your best on your own?

The H(app)athon Project

There's a growing movement to standardize the metrics around well-being that can lead to happiness. The combination of Big Data, your social graph, and artificial intelligence means everyone will soon be able to measure individual progress toward well-being, set against the backdrop of all humanity's pursuit to do the same. In the near future, our virtual identity will be easily visible by emerging technology like Google's Project Glass and our actions will be just as trackable as our influence. We have two choices in this virtual arena: Work to increase the well-being of others and the world, or create a hierarchy of influence based largely on popularity.[17]

—JOHN C. HAVENS

I wrote this article four months before I founded the H(app)athon Project. The piece was the inspiration for this book and the project I'm focusing on full-time. I believe mobile technology, utilizing sensors, will transform the world for good if personal data is managed effectively and people utilize these tools however works best for them. I'm writing this not to pitch you on the Project (although we are a nonprofit and all of our tools are free anyway, so of course I'd love for you to check it out), but because I want to be accountable to you as a writer. I love researching and writing, I love interviewing experts and providing a unique perspective. But I also believe taking action based on your passions is of paramount importance to best encourage others. That way, expertise is tempered and shaped by experience.

I was recently interviewed by the good folks at Sustainable Brands about our Project, as I spoke at their upcoming conference. Here is a description of the H(app)athon Project as I related it to author Bart King:

> Our vision is that mobile sensors and other technologies should be utilized to identify what brings people meaning in their lives. We've created a survey that's complemented by tools that track action and behavior in a private data environment. By analyzing a person's answers and data, we create their personal happiness indicator (PHI) score, a representation of their core strengths versus a numbered metric.
>
> A person can then be matched to organizations that reflect their PHI score in a form of data-driven micro-volunteerism. There's a great deal of science documenting that action and altruism increase happiness. So we're simply identifying where people already find meaning and help them find ways to get happier while helping others. At scale, we feel this is the way we save the world.
>
> At the moment, we're just beginning our work. Our survey can be taken online and on iPhones, but we're seeking funding to build out the sensor portion of our data collection. We've partnered with the City of Somerville, MA, to pilot our proof-of-concept model over the next ten months.

Somerville is the only American city to implement Happiness Indicator metrics with a sitting government. Our hope is that by adding sensor data into the mix we can gain critical insights to help with transparent city planning that improves citizens' well-being.[18]

I want to make it clear how important Hacking H(app)iness is to my life in the form of this book and the H(app)athon Project. The technology of sensors provides a way to reveal aspects of ourselves we may not see. In the same way that you achieve catharsis watching a play where actors exhibit emotion you may not always be able to reveal, sensors give you permission to act on the data driving your life.

Does your heart rate increase when you think about playing guitar in a band? Maybe you should act on that. Does your stress level increase at your job no matter what task you're doing? Could be time to switch divisions or look for new work. Affective sensors and their complementary technologies will begin to work their magic on your life in the near future if you let them. In the case of the H(app)athon Project, global Happiness Indicator metrics also provide the framework of a positive vision for the future not dependent solely on influence or wealth.

Need help defining your own vision? Check in with your data and see what revelations you have to offer yourself.

5

QUANTIFIED SELF

Wearable computing devices are projected to explode in popularity over the next year and, with a wave of new gadgets set to hit the consumer market, could soon become the norm for most people within five years. ABI Research forecasts the wearable computing device market will grow to 485 million annual device shipments by 2018.[1]

ABI RESEARCH

WE ALL CURRENTLY have wearable computing devices in the form of our smartphones. Slap a piece of Velcro on your iPhone and wear it on a headband and you're good to go. In terms of history, if you wore an abacus on a necklace back in the day, you'd also technically be part of the wearable computing movement.

In a similar fashion, quantified self as a practice has been happening since time began. When Eve asked Adam, "Does this fig leaf make me look fat?" she was comparing herself to a previous measurement. If you've used pencil and paper to figure out your finances, that form of self-tracking also fits the bill (pun intended).

Quantified self (QS) is a term coined by Kevin Kelly and Gary Wolf of *Wired* in 2007. It refers to the idea of self-tracking, or "lifelogging," as well as the organization by the same name that helps coordinate hundreds of global meet-up groups around the world. According to the group's website, the community offers "a place for people interested in self-tracking to gather, share knowledge and experiences, and discover resources." Wolf wrote a defining piece about the notion of QS in the *New York Times Magazine* in 2010. In "The Data-Driven Life," Wolf describes how improving efficiency is not the primary goal for self-trackers, as efficiency for an activity requires having a predetermined goal. Trackers pursue insights based on data collected in real time, where more questions may develop as part of an overall self-tracking process.[2]

This notion of collecting data with an unknown goal strikes most non-trackers as odd. In a world that typically rewards productivity above all, how could someone spend so much time measuring his or her actions with no set goal in mind? As with data scientists, self-trackers look for patterns in their actions to form insights versus approaching the data with hypotheses that could color the outcome of their findings.

Measuring your actions without a set goal in mind is hard. We're trained to think that all of our actions must have a defined purpose resulting in improved productivity. I remember years ago working in a high-end café and getting admonished by my manager because she felt I was moving too slowly. She taught me how to look around the café and quickly assess multiple tasks that needed to be done based on walking clockwise around the room. The lesson stuck with me. To this day I still use this technique in my own kitchen, although the only patrons I need to take care of are my kids getting ready for school.

The downside of this type of harried productivity, however, comes in the toll it can take on your psyche. It's very difficult not to gauge your success as defined by others, and the plethora of self-help guides touting increased productivity only adds to the stress.

We're coming into a time, however, when the aggregation of our data will help us automatically become more productive. Analyzing patterns and offering recommendations based on behavior provide a huge increase in productivity and value via personalized algorithms (predictive computer equations based on past actions). Stephen Wolfram, a complexity theorist and CEO of Wolfram/ Alpha, notes in an interview with *MIT Technology Review* that he stopped answering group e-mails in the morning because data showed the majority of the issues worked themselves out by the afternoon.[3] Sound familiar?

Another key benefit Wolfram describes in the article is the idea of augmented memory, when the aggregate data of our lives will be made available to us at all times. Think of your life as if every word and action were an e-mail stored in a database, searchable in an instant—that's the idea of augmented memory. The paradigm shift of fully augmented memory will have massive cultural repercussions, both positive and negative. Recording the experiences of our lives in photos and audio or video formats has been limited technically to this point due to battery life of hardware and lack of storage for content. Battery life is improving at a rapid rate, and the evolution of cloud computing (servers that access and store your data remotely, versus being stored on your hard drive) means if we can afford to pay for storage, it's available. Augmented memory enabled by these technologies will provide for the following types of applications:

- You're at a conference and someone you don't recognize smiles and walks toward you. Using facial recognition technology, you can quickly scan past life-recordings to see how you know the person.
- You can run tests on your e-mails for the past year for keystroke data (how hard you hit the keys, serving as a proxy for anger/stress) and see what times of the day or week you tend to be emotional and how that affects people's responses to your messages.

- You can cross-reference your GPS data with your e-mails, using sentiment analysis (technology that identifies certain words that infer positive, negative, or neutral language patterns) to identify the places where you are most productive.

These examples should show you why quantified self–analysis won't stay only in the realm of life-loggers or health enthusiasts for long. Hacking H(app)iness, or owning your data in this context, isn't just about protecting it—it's about liberating it to be useful in ways it's never been used before.

Objectivity in Action

Another aspect of self-measurement that's a challenge for people is staying objective. There's deep emotion tied to something like losing weight or keeping your house clean. But a key component to quantified self is the skill of articulated observation. I developed this skill over the years as a professional actor and writer. It takes practice to look at a person (or yourself) and simply record what you see. You would think stillness would be easy to achieve, but it's actually very challenging. We are hard-coded toward bias and judging others. It's in our DNA as a remnant from our ancient past when we relied on our fight-or-flight mechanisms to keep us safe.

Here's an exercise you can try to cultivate your nonjudgmental observation skills. Record yourself on video standing and reading a passage of poetry or a passage of a play. Something you're passionate about. Perform it. Have fun doing it and don't worry about the caliber of your acting. The focus of the exercise is actually about your response to watching the video. If you cringe watching a recording of yourself, pretend you're watching someone else and just describe what you see.

Most young actors (myself included) don't realize how much energy is stored in their bodies that comes out when they recite a passage of a script until the first time they see themselves on video. For instance, "flappy hand" is a common occurrence with young actors: While saying lines from a scene, their whole body will remain unmoving but one hand will gesticulate wildly as if it's caught on fire. A good acting teacher will point out the latent energy in the person's hand and have the actor take a deep breath from their diaphragm (the power center for breathing versus your lungs/shoulders) before starting again. Typically after two or three repetitions of this exercise, "flappy hand" goes away and the actor delivers a more centered and powerful performance than before.

Observation is a powerful tool. A primary trait of a gifted actor is their well-honed ability to observe humans in action. As a young actor, you "play" a character—you want a quick laugh or to milk a dramatic scene and you try to coax a certain response from your audience. That's death, because it's fake. For instance, in a comedy, characters don't think they're funny. The audience laughs because they identify with the people in the play. A good actor will *inhabit* a character, without judging the person they're playing and planning for a certain response. As a performer, they may genuinely feel terror in a role while the audience howls with delight.

Here's a story along these lines from the world of acting. You've heard about Method actors so caught up in their roles that they fully believe they've become another person. That can happen, but these stories tend to be overblown. As a professional actor, you've got to show up for eight shows a week in theater or hit your mark in film or TV. It's great to be Method and passionate, but if you lose touch with reality, you won't continue to get hired. This pragmatic aspect of performance also relates to the idea of playing an action in a scene. For instance, you can't "play" being sad in a scene— sadness is the *result* of not getting something you're pursuing.

There's a famous story of the renowned acting teacher and Moscow Art Theatre founder Constantin Stanislavski working with a group of young performers, teaching them the importance of playing an action in a scene. He asked one of his students to go onstage and sit in an armchair he'd placed there. Given no specific instruction, the young man sat and proceeded to make a series of faces that initially amused his classmates. As time wore on, however, the boy became flustered, unsure of what to do with himself. Nervous laughter from his friends faded into a tense silence. Stanislavski remained unmoving, watching with the rest of the class as a palpable sense of desperation exuded from the stage. After several more minutes, Stanislavski finally told the student he could sit down. The young man leapt back toward his seat, visibly relieved.

Then Stanislavski stopped. "Wait," he said. "I seem to have misplaced my glasses. I believe they're under the chair. Would you get them before sitting down?" The boy obliged, dropping to his knees and reaching under the chair. Then he removed the cushion, carefully examining to see if he'd inadvertently crushed the glasses by mistake. He continued looking for a few moments before Stanislavski spoke and said he realized his glasses had been in his pocket the whole time.

When the student sat down, Stanislavski revealed that the entire time the boy had been onstage, before and after looking for the glasses, had been a lesson in acting. When the boy pantomimed for his friends, he was going for an *effect*. When he was looking for Stanislavski's glasses, he was pursuing an *action*. As Stanislavski noted to the class, when the student was actively engaged in trying to accomplish a goal, however mundane, he was riveting to watch.

This story illustrates a simple fact: Truth is revealed by action. The boy was initially uncomfortable because he was trying to fabricate an experience for his friends. This same principle applies to our lives and quantifying our actions. By taking action and measuring our data without judgment, we gain insights about our behavior we didn't even necessarily set out to study.

But you won't know until you start to measure.

The Numbers on the Numbers

The Pew Internet & American Life Project released the first national (U.S.) survey measuring health data tracking in their *Tracking for Health* report. Here are some of their top findings:

- 46 percent of trackers say that this activity [self-tracking] has changed their overall approach to maintaining their health or the health of someone for whom they provide care.
- 40 percent of trackers say it has led them to ask a doctor new questions or to get a second opinion from another doctor.
- 34 percent of trackers say it has affected a decision about how to treat an illness or condition.

Particularly interesting is information about how people handled their tracking:

Their tracking is often informal:

- 49 percent of trackers say they keep track of progress "in their heads."
- 34 percent say they track the data on paper, like in a notebook or journal.
- 21 percent say they use some form of technology to track their health data, such as a spreadsheet, website, app, or device.[4]

These are some powerful statistics. Almost half of the people measured say self-tracking has changed their overall approach to help. That's huge. While they track progress in their heads, the growing popularity of quantified self apps means technology will soon increase the number of people measuring and improving their health. Hacking H(app)iness in the form of health data is a great place to start seeing how valuable it can be to measure and

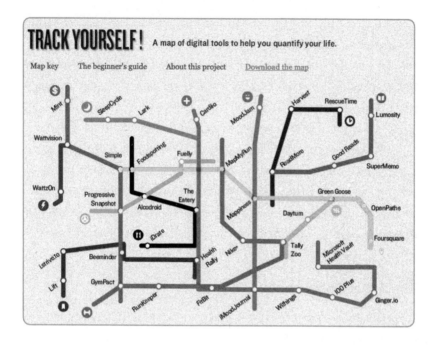

optimize your life. However, QS goes beyond measuring just the physical components of your health data.

This Track Yourself image was created by Rachelle DiGregorio[5] as part of her Track Yourself! project. Modeled after the London Underground, the different lines represent different verticals, or areas each quantified self app focuses on. The map provides a great visualization to remind the viewer how all of our behavior intersects at various points in our lives.

To get started on your own self-tracking journey, I recommend you check out the Quantified Self's Guide to Self-Tracking Tools.[6] At the time of writing, it contained 505 QS apps focused on everything from health and medicine to money and mood. Two of my favorite mood-focused apps are MoodPanda and MoodScope. MoodPanda is extremely popular, with over a million registered users regularly recording their emotions. MoodScope adds the unique and compelling feature of e-mailing your self-rated moods to friends in a model based on the "sponsor" model from Alcoholics Anonymous.

The first time I tracked behavior it was definitely empowering. It's a digital declaration of sorts. But unlike a New Year's resolution, you don't have to feel crappy after you bail in four days because you're looking for patterns that lead to bigger insights than "I like to eat a lot of bacon."

If you want to get started with a great free tool to track yourself, try iDoneThis, designed for teams in a workplace. Simply write down what you did during the day in an e-mail that's sent to you at six p.m. every evening. The company is focused on building software "you don't have to remember to use" and lets you keep a digital diary that validates your actions or those of your team.

The Art of Doing Less

Ari Meisel's life represents one of the most powerful stories I know in terms of applied life tracking. In 2007, he was diagnosed with Crohn's disease, an incurable disease of the digestive tract. After hitting a low point in the hospital dealing wih multiple medications and discouraging results, Ari began focusing on improving his health with a combination of yoga, nutrition, and exercise. He also began optimizing, automating, and outsourcing daily activities to provide himself the time he needed to get healthy. Fast-forward a few years, and Ari was declared free of all traces of what is considered to be an incurable disease. He even competed in an Ironman competition in France in 2011. It's a truly inspirational story.

Now Ari focuses on "Achievement Architecture," working with clients to help them emulate his ability to optimize in various parts of their lives, something he calls the Art of Less Doing. Here's an excerpt from one of his popular posts, "Don't Try to Prioritize, Work on Your Timing," that provides a great lesson for anyone thinking of how to track without the baggage of self-judgment:

> *Carpe Diem:* This famous saying about seizing the day is actually part of a longer phrase, *Carpe Diem Quam Minimum*

Credula Postero, which means *Seize the Day and Put Minimum Faith in the Future*. There will always be more tasks and more things that you need to get done. It would be foolish to think that simply arranging tasks in a pecking order will have any bearing on your productivity or your life tomorrow or even an hour from now . . . When it comes down to actually getting things done, we must live in the moment. You assign yourself a task at the relevant moment, you complete the task, and you move on. You don't worry about what you have to do next because your system will "assign" it to you when the time comes. You are delegating the responsibility of worrying about these things to a system of productivity, the system of Less Doing.[7]

Technology can help us live more in the moment. Quantified self tools provide a path to help us study ourselves so, after we gain an insight about our behavior, we can optimize accordingly. We just have to get out of our own way to let it happen.

H(app)y and Healthy

As a way to provide more examples on how quantified self tools and other methodologies can increase your happiness and well-being, I've included an article I wrote for Mashable called "How Big Data Can Make Us Happier and Healthier."[8] My hope is that it will provide you with a number of ways you can Get H(app)y in your life.

HOW BIG DATA CAN MAKE US HAPPIER AND HEALTHIER

Big Data is getting personal. People around the globe are monitoring everything from their health, sleep patterns, sex, and even toilet habits with articulate detail, aided by mobile technology. Whether users track behavior actively by entering data or passively via sensors and apps, the quantified self (QS) movement has grown to become a global phenomenon, where impassioned users seek context from their Big Data identities.

Moreover, with services like Saga and Open Sen.se, users can combine multiple streams of data to create insights that inspire broader behavior change than by analyzing a single trait. This reflects a mixed approach design (MAD) research methodology that purposely blends quantitative and qualitative factors in a framework where numbers are driven by nuance. The science of happiness, for example, is now a serious study for business, as organizations combine insights of the head and heart to create environments where workers feel their efforts foster meaningful change.

However it's studied, the desire to understand monitored behavior has reached a fever pitch, and the QS movement is attempting to meaningfully interpret our daily data.

The Power of Passivity

"We are moving towards a time when the ability to track and understand data is deeply woven into our daily lives," says Ernesto Ramirez, community organizer for Quantified Self. "Sensors are becoming cheaper and connectivity is more ubiquitous by the day."

This ever-present nature of data availability will become even more powerful when the general public begins to use apps that require little ongoing attention or input. Passive data collection is especially relevant in the health-care industry, for example.

"The data Quantified Self provides is not a replacement of any measurement to date—we haven't had this type of measurement to date," says Halle Tecco, cofounder and CEO of Rockhealth, the first seed accelerator for digital health startups. "Patients live very cautiously before trips to doctors, and

this causes more trips to doctors. It's better if physicians can get a more comprehensive view of people's ongoing health."

Tecco highlights the importance of passive monitoring. For instance, a mobile app can continuously measure glucose levels or other factors like heart rate over time. Spikes in those readings could immediately trigger a doctor, even remotely. "We can save money and improve outcomes by having data collection embedded in our everyday lives," she adds.

Declaring Your Deeds

Nowadays, people are declaring their daily goals and intentions to peers and seeking their support via social media. Companies like Gympact and StickK operate on accountability-based influence (ABI), a scenario in which you're judged on your actions versus your words. Beeminder, a "motivational tool that puts your money where your mouth is," falls into this category too, according to cofounders Daniel Reeves and Bethany Soule. Users quantify a goal and pledge to pay money to Beeminder if they fall off the wagon.

"The platform lets users tweak their regimen at any time, with the caveat that any changes take effect with a one-week delay," says Reeves, "so you can change your commitment, but you can't change it out of laziness, unless you're particularly forward-thinking about your laziness."

According to Reeves and Soule, Beeminder is the only platform that combines the advantages of quantified self-tracking with a commitment contract, a compelling and self-binding form of digital declaration in which users risk a public pledge as a form of accountability for their goals.

Other companies in the QS space offer tangible ways to demonstrate action. A simple framework for tracking positive behavior is provided by uGooder wherein users gain

badges for broadcasting good deeds they've completed. The service also lets users print a transcript of all the good deeds they've ever done using the platform.

"I thought someday this might be something people could take to a job interview or submit with a college application to show how much good they have done," says Dan Lowe, uGooder's creator. The idea is compelling—why shouldn't employers or schools focus on overtly positive, community-supported behavior, versus an errant photo of high school revelry?

The rise of portfolio platforms like Pathbrite and Linked-In's volunteer profiles encourages people to professionally self-claim their positive behavior. The rise in ABI will eventually supplant trust networks built primarily on words.

The Advantage of Aggregation

"We are of the philosophy that data is versatile," says Rafi Haladjian, cofounder of Sen.se. "Once you collect data from a source, you can decide how to use it later on."

Haladjian seems more artist than engineer. He credits the muse of serendipity for guiding data in ways that maximize insights for enlightened users. Sen.se also proselytizes the "Internet of Everything" over the Internet of Things, supporting the idea of the interconnectivity of data when multiple passive sensors work in unison, versus one input alone.

"We need to create the culture of data mashups and we're finding ways to make that easier," he says. Demonstrating how to identify unique patterns via these mashups, Haladjian speaks of an elderly parent whose passive sensor placed in her favorite armchair measures how much time she spends sitting. The sensor is one of many placed throughout her home to gauge time spent in various locations or usage

of different appliances, data the woman's caretaker can use to measure her health.

In this instance, information is collected without its full purpose known beforehand. "If users start to simply collect data in this way," notes Haladjian, "they can use all sorts of tools to discover the hidden meanings that lie behind the mundane."

Esther Dyson, chairman of EDventure Holdings, also studies the concept of data mashing. Her concept of the quantified community interprets Big Data as a series of inputs, driven by individuals who wish to improve their communities and world. She describes her vision of quantified community in an article for Project Syndicate: "I predict (and am trying to foster) the emergence of a quantified community movement, with communities measuring the state, health and activities of their people and institutions, thereby improving them."

For example, she says, when QS tools collect data about health, this data can and should be combined with local health statistics to generate new insights. She also notes the existence of civically minded apps like Street Bump that let users take photographs of or collect data around potholes or other citizen concerns.

This community focus shows how the QS movement can provide a new layer of qualitative data on top of quantified reporting. Think about an app wherein citizens could report their emotional state at seeing a pothole, as well as record its location. QS apps could easily aggregate these emotional tags with obvious economic repercussions. (If you look for good schools when buying a house, wouldn't you also check the "emotional history" of a neighborhood as well?) Combine this tagging with the ability to search the virtual arena via augmented reality tools like Google Glass and it's easy to see how the quantified community will usher in a transformative era of civic engagement.

Emotions in the Enterprise

"Altruism is alive and well on the Internet," says Paul Marcum, director of global digital marketing and programming for GE and a driver of Healthy Share, a Facebook app that lets users announce health goals and use friends as sources of inspiration. "There is an opportunity to have users 'pay it forward' when they build themselves up by helping others," he says.

The platform proves that the idea of quantified self has taken hold with brands and enterprise. Marcum points out that "sharing is a form of tracking," that announcing actions via social media is akin to active monitoring via a QS device. "This is information people want to share, and we want to know how to capture that to spark behavior change," says Marcum. Platforms wherein users are driven by intrinsic motivation and supported by a community let brands get out of the way and understand what truly drives a user base.

"Why is ethical integrity, why is character, not considered an economic asset in a time when trust and reputation are widely heralded as competitive advantages for companies?" asks Tim Leberecht, chief marketing officer of NBBJ, a global design and architecture firm, who while still at his previous role at Frog Design was a driver for the company's Reinvent Business hackathon, an event to "create concepts and prototypes to help create a more social and human enterprise."

In a post titled "Hope for the Quantified Self," he refers to mounting evidence that shows well-being and happiness increase productivity and the bottom line. The result is organizations seeking to understand what truly makes employees happy, how to best blend qualitative along with quantitative metrics, a practice that may seem foreign to most corporate cultures.

Leberecht has a solution: "We need to find a way to measure the social value created by those whose contributions are outside of the common ROI vocabulary."

He cites the CEO who inspires and instills hope for thousands of employees, but who has failed to meet the board's growth expectations. "As hyper-connectivity and social networks tear down the boundaries between professional and private lives, only those who are complete will be able to compete." Matching internal and external character, words and deeds, these new "whole selves" will no longer tolerate a chasm between idealism and pragmatism.

Leberecht's observations point to a growing pressure for organizations to study happiness within the workplace. Corporate restrictions may soon lift to proactively embrace character traits from outside the workplace, and qualitative paradigms will gain the credibility of quantitative metrics.

"What if we were able to take the quantified use of metadata, a computing-based narrative of humanity, and integrate it with centuries of human narrative and storytelling?" asks Thanassis Rikakis, vice provost for design, arts and technology at Carnegie Mellon University. "That would provide a tremendous opportunity to understand humanity at a level that's never been understood before."

Rikakis is the founder of Emerge, an event that first took place at Arizona State University. Featuring noted science fiction writer Neal Stephenson and visionary geek Bruce Sterling, the event also brought together scientists, artists and designers. The primary goal was to bring together experts from multiple disciplines, recognizing that purely quantitative solutions can't fully tackle the complex issues we're faced with in the modern era.

Rikakis points out that QS technology allows us for the first time in human history to embed computing in every part of our lives. The value of the quantified self will be amplified when we recognize how qualitative measures complement Big Data.

For his work in interactive neurorehabilitation, Rikakis uses highly advanced tools to track forty-four kinematic parameters of the affected upper limbs of stroke survivors. He says data from these kinematic measures and their relation to functional outcomes is an essential step towards promoting recovery. But to be effective, this data needs to be combined with the qualitative observations of a therapist and filtered through the relationship of therapist-patient.

"We have to keep in mind that there's information that does not go through data but via human interaction," says Rikakis. "It goes from community to community and has a richness that's hard to quantify."

Don't Worry, Be App-y

We're in an era when sensor technology and the maturation of smartphones mean data is being collected about your actions in ways that have never existed before. There are no universal privacy and identity standards, which means your unwilling contributions to Big Data are being shaped by forces you can't control.

The good news: Getting familiar with quantified self applications will benefit personal and community self-awareness. You'll understand how to better shape your identity in this new virtual economy and learn the *quantitative* metrics that derive their fullest context when seen through a *qualitative* lens.

6

THE INTERNET OF THINGS

I believe that the Internet of Things can be that powerful lever to create ground truth with the right data; to engender more trust among people and institutions and to elevate time as the most important outcome of our efforts. [The Internet of Things] will actualize the great synthesis between people and machines; analog and digital; silicon and carbon.[1]

CHRIS REZENDES

THERE ARE A LOT of things we don't see that affect us. If you have a less-than-developed sense of smell, you might not recognize that your idling car is kicking out some noxious fumes that can negatively affect your health. Likewise, you might not realize when a tree in front of your house is producing more oxygen at a certain time of year.

The Internet of Things (IOT), sometimes referred to as the Internet of Everything, refers to technology embedded in objects around us that has become inexpensive and ubiquitous enough to record or broadcast data on a regular or real-time basis. You're familiar with these technologies in bar codes on the products you buy—these are used in a supply chain process to help the people who make specific items get to where people buy them.

You may have heard of another technology called radio-frequency identification (RFID) that many see as the precursor to the Internet of Things. RFID tags have been used in supply chain logistics for years. Tags emit a low-frequency signal containing information about the contents of a package or other container. This quickly allows a worker with a tag-reader (an electronic wand like you'd see someone using at the supermarket) to scan all the boxes in a truck to know their contents without having to open them up.

The phrase "Connected World" is not just a metaphor. We are becoming more connected to and through the things around us on a daily basis. How we feel about ourselves and other people is deeply affected by how we interact with our surroundings. Now the world around us can more deeply interact with itself with or without our involvement.

Let's take the tree in front of your house I mentioned. Hearing that it produces more oxygen at certain points of the year may be interesting but not terribly relevant information. But what if you knew that trees also lower cortisol levels, a primary contributor to stress, plus they can remove negative pollution? As reported by Anne Hart in the *Examiner* article "How Trees Contribute to Health by Producing Oxygen and Lowering Cortisol Levels," "the urban trees of the Greater London Authority (GLA) area remove some-where between 850 and 2,000 tons of particulate pollution (PM10) from the air every year."[2] The article goes on to explain how this data may affect urban planning so trees can be planted between highways and nearby schools and homes.

An article from *Smithsonian* magazine, "Going to the Park May Make Your Life Better" by Sarah Zielinski, describes some more findings from a report by the National Recreation and Park Association relating to trees and health, including the fact that in Los Angeles, people who had more access to parks reported a higher level of trust for people in their community, or that children with attention deficit disorder had better concentration after walking in the park than in an urban setting.[3]

Why these reports have such impact regarding the Internet of Things has to do with the pragmatic impact data from trees and the environment will have on our lives. In the near future, planting a tree in a low-income area may be a primary tactic to curb violence. Eventually data might show how strategic greenery directly correlates to fewer hospital visits for local residents, lowering health costs. On an individual level, if you suffer from high stress levels according to your wearable device, you may get a text from the tree in your front yard saying, "Come sit by me for ten minutes— your cortisol levels are through the roof!"

It's this type of deep connectivity that is being empowered by the technologies composing the Internet of Things with what Chris Rezendes, president of INEX Advisors, refers to by his idea of "ground truth." Objective data from sources we couldn't unlock in the past will inform and shape our lives in ways not possible before.

Machine-to-Machine Mentality

In a broad definition of the Internet of Things, machine-to-machine (M2M) technology refers to how devices communicate with one another. Sometimes called peer-to-peer (P2P) networks, these technologies form the backbone for interoperability along the things that surround us in our lives, including our cars and homes. These technologies are also already here—you've likely gotten Wi-Fi for your computer using a local area network (LAN), for instance. The combination of these technologies allow for varied applications of IOT that help demonstrate its growing ubiquity.

I had the pleasure of interviewing Vint Cerf, VP and chief Internet evangelist at Google, most widely known for being one of the inventors of the Internet, for my Mashable article "The Impending Social Consequences of Augmented Reality." Regarding the Internet of Things, he noted that the ability to monitor data on a twenty-four-hour basis would greatly help us quantify our understanding of the world. One specific business example he provides, related to wineries, wasn't included in my Mashable piece, but I wanted to add it here:

> With GPS receivers, winery owners are beginning to monitor what nutrients each plant needs to maximize productivity of each wine. Instead of analyzing the average output of a vineyard, owners can measure productivity on a plant-by-plant basis. This helps them maximize their yield or optimize the quality of specific fruits by caring for plants in different ways according to data. This is an example of how computing power, memory, and local miniaturization are enabling things we couldn't do before.[4]

Keeping with the wine theme, Cerf went on to describe how apps could start to recognize if our blood alcohol content (BAC) is too high. Embedded with this technology (like the Last Call app that predicts when your BAC will peak), someday our cars won't start when we turn the key after sensing our inebriated breath and will say, "Had a few too many there, sport—just called you a cab." According to the Centers for Disease Control, in 2010 over 1.4 million drivers were arrested for driving under the influence of alcohol, and intoxicated driving resulted in over ten thousand deaths.[5] Uses of technology like this scenario could save lives, lower insurance rates, and allow police officers to spend time on other areas of need rather than on highway patrol—all by utilizing the data created by our interaction with the Internet of Things.

"Dishwashers and X-ray machines. In some shape or form, they're computers." M. Mobeen Khan is executive director of Mobility Marketing for AT&T Business and told me the following in our interview for this book:

There is a lot of data in these appliances and machines, data people are not necessarily using. For our work and clients, we want to run the diagnostics of these machines on a common platform where data can be better analyzed. A typical X-ray machine may be used over twenty times in an hour. If I can analyze data about what that machine did over the course of a year, I can get a richer sense of how people are using it to improve future designs of the product.[6]

Bill Zujewski, CMO and EVP of product strategy for the Axeda Corporation, reveals similar benefits of IOT technology for his clients, something the company calls "connected capabilities." Like Khan, he notes how dishwashers, outfitted with firmware allowing operating systems to be connected to a manufacturer, could be repaired remotely versus having to be recalled when damaged.

In our interview for this book, Zujewski also likens the evolving world of IOT to the evolution of the app economy:

This is where machines are going. Pretty soon you're going to get a coffeemaker and wonder why you can't program it from your phone. The precedent set by Apple and Samsung around apps will spill into our lives regarding our appliances and other machines.[7]

The app logic will apply to both consumers and the business world, as Zujewski noted with an example from a client, the Getinge Group, an organization focused on providing sterilization and other contamination services. Getinge worked with a hospital client that utilized large commercial dishwashers requiring workers to monitor equipment around the clock. Ninety-nine percent of the time, the washers operated without a problem, but the workers still had to monitor equipment on-site for the one percent of the time a problem could hinder cleaning. Getinge provided an app that could monitor and even restart equipment remotely, allowing workers to go home and spend more time with their families.

Ground truth and smart Internet of Things applications are improving our lives and work.

"If my TV speaks AllJoyn, my washer can tell me when it's done with a load by sending a message I'll see on my screen." Liat Ben-Zur is a senior director of product management at Qualcomm and leads the AllJoyn business, focused on the company's Internet of Everything software strategy. While many Internet of Things technologies rely on Wi-Fi connectivity, where data is transferred or stored in the cloud, AllJoyn is a proximity-based network—data can privately pass between two devices. Here's how she explained this in our interview for this book:

> As devices and appliances get connected and smart, where does privacy come in? If I have a connected garage-door opener, do I want a manufacturer in the cloud to know every time I'm coming into and out of my home? In the future, people will only offer up their data when someone solves a problem or adds value to their lives. This is one of the main reasons proximity networks offer an untapped resource—people can engage with the world around them without super-private data being exposed.[8]

H(app)iness in Everything

Technology research company Gartner named the Internet of Things among its Top Ten Strategic Technology Trends for 2013,[9] reporting that more than thirty billion objects will be connected by 2020. It's going to become increasingly difficult to find places that aren't part of the Connected World.

This means we're going to have dozens of new ways to measure our emotions and well-being. Even without being an active self-tracker, you may buy a Nest smart thermostat that learns your temperature preferences and makes you happy by lowering utility bills when it turns down the heat by itself when you're away from home. Or you may get a smart fridge from Samsung that offers recipe suggestions based on the food you have in the freezer. In the future this type of fridge may e-mail FreshDirect once a week, replenishing items set to expire, but only if they synch with your nutrition regimen as recorded on your Weight Watchers app.

Our health and happiness will be even more tied to technology than they are right now. And where we save time or energy utilizing the Internet of Things, we'll be able to improve our well-being with a record of how we optimized in most every situation.

But ethical and privacy issues will increase as IOT becomes ubiquitous as well. Perhaps boxes in the future will be outfitted with pressure sensor tagging as well as RFID sensors. Designed to analyze how customers open packaging to improve future designs, these test tags may also identify and record the people who smashed boxes with their fists because they couldn't get them open. This might reflect in an accountability score of some kind that could be reflected in an identity score others could see. That example probably wouldn't hurt your chance at getting a date, but you wouldn't get a job at the post office.

Hacking H(app)iness will require balance as we move toward the future. Part of our ground truth will be learning how to *stay* grounded within the boundaries of these amazing technologies.

ARTIFICIAL INTELLIGENCE

Welcome to the next phase of computing ... Companies ranging from IBM to Google to Microsoft are racing to combine natural language processing with huge Big Data systems in the cloud that we can access from anywhere. These systems will know us better than our best friends, but will also be connected to the entire Web of Things as well as the collective sum of all human knowledge.[1]

GREG SATELL

WE ARE ALL creatures of habit. We are steeped in binary behavior, making choices between two options that will lead to a desired outcome. Here's an example of this from my own life in my morning routine:

- Start coffee, unless dog is barking loud enough to wake my wife.
  - If dog is barking, take him outside to pee, and then make coffee.
- If dog is not barking, smile at him and give him peanut butter.
- Pour soy milk in my mug and cream in my wife's mug while waiting for coffee.
  - If there's no soy milk, curse, then pour a small portion of cream in my mug.

- Savor first sip of coffee after *first* handing other mug to wife.
  - If interrupted by children, unless they're on fire, finish coffee.
  - If children are on fire, curse, then put them out before reheating coffee.

Years ago I read an excellent book called *The Weekend Novelist* by Robert J. Ray, which provides a fifty-two-step plan to help you write a book in a year. He made an excellent observation about human behavior, which is that we're "mired in ritual." We get up every morning and do the exact same things—pee, take a shower, shave. When I'm on the road for work, I organize my rituals in such a way that I can actually look forward to traveling.

We think like machines. We also think like humans, but decision-making based on predetermined outcomes is a part of our lives.

Take dating, for instance. Most people feel they have a type of person they'd like to have as a partner. But are the criteria we think will make us happy in another person always right? Speaking for myself, before I met my wife, my choices in dating *sucked*. This isn't a criticism of the women I went out with, mind you. My dating pattern before I met Stacy was to have short relationships with women I knew weren't ready for commitment. Mutual usury seemed to be working for me until the nadir of my romantic life when, during a first date, a woman told me, "Technically I'm still married," and I didn't flee. The familiarity of my dating pattern had brought such comfort, I couldn't see the damage it was doing.

Here's how Wikipedia defines an algorithm: "In mathematics and computer science, an algorithm is a step-by-step procedure for calculations. Algorithms are used for calculation, data processing, and automated reasoning."[2] In terms of dating, the use of an algorithm mentality can be found in Amy Webb's memoir, *Data, A Love Story: How I Gamed Online Dating to Meet My Match*. The book is funny and real and documents her use of data and algorithmic behavior to accurately assess her perfect man, which she does. I wish I had thought of it back with "still married" lady.

eHarmony is taking the matchmaking algorithm to a new level, according to John Tierney in the *New York Times*. While presenting research at the Society for Personality and Social Psychology, eHarmony's senior research scientist, Gian C. Gonzaga, reported, "It is possible to empirically derive a matchmaking algorithm that predicts the relationship of a couple before they ever meet."[3] While the statement drew a great deal of criticism from peers, eHarmony has gathered answers from over forty-four million people on questionnaires featuring over two hundred questions. The data set is substantial enough to warrant credibility.

Predictive algorithms tend to freak people out. We don't like to think our intentions can be gamed or guessed. We're also distrustful of how they can shape our perceptions of the world, as Eli Pariser, chief executive of Upworthy and board president of MoveOn.org, wrote about in *The Filter Bubble: How the New Personalized Web Is Changing What We Read and How We Think*:

> Left to their own devices, personalization filters serve up a kind of invisible autopropaganda, indoctrinating us with our own ideas, amplifying our desire for things that are familiar and leaving us oblivious to the dangers lurking in the dark territory of the unknown. In the filter bubble, there's less room for the chance encounters that bring insight and learning . . . If personalization is too acute, it could prevent us from coming into contact with the mindblowing, preconception-shattering experiences and ideas that change how we think about the world and ourselves.[4]

It's a point we all think about concerning our interactions with machines: Where do we lose intentionality in our actions? Does our Connected World demand the sacrifice of serendipity?

Not according to Greg Linden, former principal engineer at Amazon who invented the company's recommendation engine and personalization framework. In response to how personalization algorithms could supplant serendipity, Linden told me during an interview:

> For the early work on recommendations at Amazon, it always had the goal of helping people find books they wouldn't otherwise find. You can only search for something if you know it exists. You have to embrace serendipity to discover new things. It's more of a process of wandering than of searching. But it would take forever to wander through a five-million-item catalog. Even the earliest recommendation features at Amazon were designed with the idea of helping people wander, helping them discover things they wouldn't find on their own.[5]

"Machine learning" refers to systems that can learn from data. Artificial intelligence enlarges this idea to incorporate systems that can learn from their surroundings. Humans learn from data and their surroundings, but at different rates. We also aren't programmed to constantly monitor our lives to optimize at all times.

At least until now, because we can. Hacking H(app)iness wasn't possible before passive sensors and mobile phones became widely available, tied together by the Internet of Things. We have more opportunities to quantify our emotions than ever before, testing our perceptions and beliefs in the wake of ordered data. This process doesn't have to be scary. While I loathe the idea of usurping serendipity for technology, I still use my GPS almost daily. I've made a trade-off—I get lost less, knowing I may also never stumble upon a glorious restaurant not registered by TomTom. But I still meet people I didn't before, and they tell me about cool places they've discovered. Serendipity once removed still stimulates. Assisted wandering works for me.

We can call ourselves Luddites and say we're not on Facebook. We can point to an older relative who doesn't own a cell phone or talk about emerging countries in the world where technology doesn't exist. But of the world's estimated seven billion people, six billion have access to mobile phones,[6] and only four and a half billion people have access to working toilets. Rather than decry the use of algorithms for fear of losing serendipity, we should focus on creating technology with positive intent that can help other people in our Connected World.

Relation-chips

Robotics technology holds the potential to transform the future of the country and is expected to become as ubiquitous over the next decades as computer technology is today.[7]

—THE ROBOTICS VIRTUAL ORGANIZATION

Our future is inexorably tied to robots. Autonomous machines build our cars, perform surgery, travel to inhospitable depths of the sea, and participate in combat operations to keep human lives from being put at risk. Soon, robots will become commonplace in assisted living centers, providing companionship for the teeming number of boomers requiring specialized care. Our homes and cars already feature primitive forms of robots equipped with Wi-Fi that can communicate to manufacturers or other devices around the Connected World.

The notion of The Singularity has been popularized by Ray Kurzweil, renowned author, inventor, and director of engineering at Google. The term refers to a date in time when computer intelligence catches up to and surpasses that of human beings. While opinions vary, The Singularity is predicted to occur within twenty to forty years, probably around 2040. The logic for this assertion is based on Kurzweil's Law of Accelerating Returns, the idea that technology increases at an exponential versus a steady rate. This means it takes less time for new technological advancement to occur by utilizing tools that didn't exist before. A simple example: You build a house faster with a hammer than a rock. Once electricity was invented, pneumatic hammers shortened building times even further.

Ethics and identity in the wake of robotic technology need to be considered *now*. Living an examined life has to include individual decision on how we interact with an intimately connected world.

"Right now, engineers have to be certified to work with elevators. Soon we'll have emotional ethics certification for people dealing with technology that interfaces with our emotions."[8] Alexander Reben created the project Blabdroid, "the world's first documentary shot and directed entirely by robots," that was featured in Amsterdam and at the 2013 Tribeca International Film Festival's Storyscapes Competition. I interviewed Reben along with filmmaker Brent Hoff, who worked on Blabdroid as well as creating *The Love Competition*, a documentary featuring people competing in MRI machines to express emotion.

For both projects, Reben and Hoff were investigating the theory of the ELIZA effect, named after an experiment conducted by MIT computer scientist Joseph Weizenbaum in 1966. In the experiment, users would respond to different recordings from a simple boxlike machine (named ELIZA) programmed to play recordings based on people's verbal responses. Here's a sample conversation between ELIZA and a young woman as recorded in "Dialogues with Colorful Personalities of Early AI" by Güven Güzeldere and Stefano Franchi:

YOUNG WOMAN: Men are all alike.

ELIZA: In what way?

YOUNG WOMAN: They're always bugging us about something specific or other.

ELIZA: Can you think of a specific example?

YOUNG WOMAN: Well, my boyfriend made me come here.

ELIZA: Your boyfriend made you come here?

YOUNG WOMAN: He says I'm depressed much of the time.

ELIZA: I am sorry to hear you are depressed.

YOUNG WOMAN: It's true. I'm unhappy.

ELIZA: Do you think coming here will help you not to be unhappy?

YOUNG WOMAN: I need some help, that much seems certain.[9]

What began to occur with the machine was both remarkable and disturbing to Weizenbaum: People began to unconsciously assume ELIZA's questions implied emotional interest in their problems, even when they *consciously knew ELIZA was only a machine.* They were anthropomorphizing ELIZA and experiencing a therapeutic release, revealing problems that ostensibly weren't being listened to by their human companions.

What's so fascinating about this effect is the balance between conscious and subconscious response elicited by users. Not only did they forget they were speaking to a machine, *they didn't care.* A number of studies[10] continued to document this behavior in the decades that followed, programming robots to mimic eye or voice patterns in humans to garner sympathy and emotional response.

To study the ELIZA effect in action, Reben and Hoff created Cubie, a small cardboard robot outfitted with a camera, wheels, and a set of prerecorded questions voiced by a seven-year-old boy, including, "What's the worst thing you've done to someone?" and "Tell me something that you've never told a stranger before." Footage taken at the International Documentary Film Festival in Amsterdam revealed a number of candid responses, including one young woman's response to the worst thing she'd done to someone: "I didn't tell my father I loved him before he died."

Blabdroid isn't intended to be manipulative, however. As Hoff pointed out in our interview, the experiment is designed to provide an emotional outlet for people based on deeper questions than are addressed with modern entertainment:

> Instead of watching a reality TV show, we're interested in what kind of emotional reaction people will have with a little robot. Despite a relatively low level of artificial intelligence, people are having phenomenally emotional experiences. And isn't that the point? Do robots have to be incredibly smart to make our lives better? No, they just have to be designed right and fit.[11]

Hoff's documentary *The Love Competition* also explores the intersection of emotions and machines. Seven volunteers met with Stanford University neuroscientists who measured their brain patterns in an MRI machine. Volunteers were asked to vividly imagine their experiences with a current or past love, where a winner would be determined based on output of brain activity focused on emotion. The results are described by Angela Water-cutter in the *Wired* article "Neuroscientists Measure Brain Activity in *Love Competition*," where she points out that, based on physio-logical results (levels of dopamine and serotonin activity), people can show they love someone more deeply than someone else can.[12] Having watched the video myself, what was more powerful than the empirical evidence was the effect the experience had on com-petitors, who expressed deep emotion after leaving the MRI, many of whom were almost in tears. And in this case, the ELIZA effect of the MRI machine is less overt than with Blabdroid, but still just as poignant: People willingly, or inadvertently, will express emo-tions with the presence of robots or technology that would have stayed hidden without them. Fueled by a sense of freedom to express sentiment that may be construed as inappropriate or questionable by humans, people open up to machines. Even though they know they're doing it.

In a final insight about the nature of people's responses to artifacts engineered by humans in our interview, Reben brought up a powerful point about the nature of some of our oldest companions:

> A lot of people have fears about artificial intelligence and social robotics. They think, if I get a robotic animal as a pet, won't that be bad? I'll be replacing social connections with technology. Newsflash—we've had this precedent for eons. It's called a dog. Dogs have been technologically bred for generations through genetic selection to be our companions. Carbon or silicon, sometimes we need to vent our emotions on something that's nonjudgmental.[13]

Reflections

Mirrors aren't always fun. In light of how we're looking at ourselves, we may smile and love what we see. Or we may view ourselves through a lens of criticism, noting every blemish. Quantified self and the Internet of Things provide multiple ways to reflect on our humanity. They also let others peek from behind our shoulders and see us in ways we didn't recognize before.

Being accountable in the Connected World with its multifaceted mirrors doesn't need to be scary, just informed. The tools involved, like ELIZA, can provide catharsis versus criticism on your journey to optimization. But as the rate of technology is increasing exponentially, you can't afford to linger at the glass without embracing your digital identity. Privacy isn't dead, but requires being proactive—be accountable so the identity you broadcast is the one you mean to project.

Be a Provider

BROADCASTING VALUE IN THE PERSONAL DATA ECONOMY

I WANT YOU TO GET UP RIGHT NOW
AND GO TO THE WINDOW. OPEN IT,
AND STICK YOUR HEAD OUT, AND YELL,
"I'M AS MAD AS HELL, AND I'M NOT
GOING TO TAKE THIS ANYMORE!"
—*Howard Beale, in the film* Network

BIG DATA

Courts have recognized celebrities' claims to a property interest in their name and fame to seek compensation whenever such an image is used for a commercial purpose. Why not extend such a property interest to the personal data of ordinary individuals? For, with the advent of digital technologies, hasn't personal data of us all become an asset that is worth real money?[1]

CORIEN PRINS

THIS ISN'T A BOOK about getting angry. But it is a book about Hacking H(app)iness, which involves reevaluating ideas about the ways you measure what you value in your life. And like Howard Beale from the movie *Network*, I think your life has value. And in the Connected World, that value is *fiscal* as well as inherent.

Much of the debate around privacy with new technologies doesn't stem from ethics, but economics. When I say you have a right to privacy no matter what your preference, I'm also saying you have a right to your money. Your currency. The stuff you put in a bank.

Nonetheless, in relation to privacy issues, it's common to hear phrases like "But kids these days don't care about privacy—they're used to sharing their pictures on Facebook and grew up using social media." First off, the scope of these statements is simply absurd. Not all "kids" feel the same about privacy, plus most teens have a better awareness about setting their privacy controls than many adults. Secondly, once you become aware that data regarding people's identity is being sold, stop making the conversation about privacy. Make it about economics and see the reaction.

Old Conversation

CONCERNED ADULT: Don't you care you're giving private data away to brokers?

SAMPLE YOUTH: Not if they give me a coupon or whatever.

New Conversation

CONCERNED ADULT: Don't you care you're giving $1,200 per year away to data brokers in exchange for a few coupons?

SAMPLE YOUTH: Why don't I get any of that money?

Alexis C. Madrigal elaborated on this point in the *Atlantic*:

In a survey by Carnegie Mellon's Lorrie Cranor and Stanford's Aleecia McDonald, only 11 percent of Americans would be willing to pay a dollar per month to withhold their data from their favorite news site. However, 69 percent of Americans were not willing to accept a dollar discount on their Internet bills in exchange for allowing their data to be tracked. That is to say: if people think data is already flowing to a website, few would pay to hold it back . . . The companies making the data-tracking tools have serious incentive to erode the idea of privacy not just because they can make (more) money, but because privacy erosion leads to more privacy erosion.[2]

One of the primary reasons we've all become complacent about privacy is not just because of our preferences toward technology;

it's because the people who stand to lose money if we own our data don't want us to cut into their profits. While the value of a person's data depends on things like their age, where they live, and how much time they spend online, keep this point clear in your mind: Other people make more money off of your data than you do.

While the Internet advertising model shifts to adopt consumer awareness of the personal data economy, we also need to be accountable for our actions regarding payment of content providers. The 11 percent of Americans willing to pay one dollar to withhold their data may be opting to pay that dollar to the news site as an exchange of value. Content providers need to pay bills like anyone else, so many offer visitors the chance to pay in exchange for an advertising-free environment. We've been trained, however, to know we can find similar free content on dozens of sites, so typically don't remain loyal where content feels commoditized. By and large, this means content providers need advertising dollars to derive revenue from any eyeballs that visit their site, however fleetingly.

What this means for consumers is we want the best of both worlds: We don't want to be tracked or have our data be sold to brokers. But we're also not willing to pay for content, so we unwittingly keep a broken advertising model afloat that erodes consumer privacy while profiting a diminishing number of Internet services that don't want things to change.

But that's the way things are, you say. Who cares?

You do. You just don't realize how technologies like augmented reality and facial recognition mean people can tag your image and sell it like they do right now. But the visual economy doesn't have terms and conditions for you to sign. Whether or not you care about privacy, if the broken Internet model goes virtual, people make money off your image and identity without your even knowing. So the next time you're thinking of leaving a content provider's site because they asked you to contribute money so they don't have to be reliant on advertising dollars, remember: In the virtual world you're not just the product.

You're the content.

The Personal Data Economy

If your personal data is the same as money, it deeply affects the economics of your life. When you broadcast your data, whether it's personal via quantified self or public via the Internet of Things, start picturing yourself walking around with dollar bills hanging out of your pockets. Then picture someone taking those dollars from your pockets while saying, "Can you wear looser pants tomorrow to make it easier for me to fleece you?"

Seriously, picture this image and tell me you're still complacent. As a parent, picture someone doing that to your kids while they're also exposing their image and private information for anyone to see or stalk. Are you angry yet? Now picture a future where this practice will accelerate, where your currency gets traded without your involvement. Now join me in opening a window to let everyone hear you as you scream to the world, "I'm mad as hell, and I'm not going to take this anymore!"

People controlling an economy control power. The term "Big Data" could just as well be called "big money." It doesn't make a difference if your data may not be worth as much as someone else's. It's worth *something*. When broadcast, any digital information relating to your image, words, or actions becomes part of the personal data economy. Broadcasting in this context can be viewed in an economic context, where personal data is an issue of property versus privacy. As Corien Prins notes in an article in *SCRIPTed*:

> In looking at privacy as a problem of social cost, commentators have argued that the prospects for effective personal data protection may be enhanced by recognizing a property right of such data. They feel that the present conception of

privacy is an ineffectual paradigm and that, if we want strong privacy protection, we must replace it with the more powerful instrument of a property right.[3]

Data is property. Intellectual property, or IP, is such a big deal for companies because it implies ownership. Your likeness, actions, and history belong to you—at least until you give them away.

You've got an intimate and personal stake in the Big Data revolution: Your Little Data is part of it.

The Basics of Big Data

Among data scientists and tech geeks, "Big Data" is largely seen as a marketing term. It's too general a concept to be tied to any one industry, referring to the concept of massive data sets from multiple sources. The term evolved as chips and hardware have become smaller and cheaper in the past few years, allowing for an explosion of data that currently has been too granular to collect. To give a sense of how vast the world of Big Data has become, the International Data Corporation's digital universe study from December 2012 says that the digital world will reach forty zigabytes by 2020, an amount that is equal to fifty-seven times the amount of all the grains of sand on all the beaches on earth.[4] You begin to see why using the phrase "Big Data" is akin to saying "the Internet." It's too vast and general to have meaning in and of itself. As Rufus Pollock, founder and codirector of the Open Knowledge Foundation, pointed out in his article for the *Guardian*, the size of data isn't the priority. What's important is having the data and the best context to derive insights from it.[5]

In terms of characterizing Big Data, many experts refer to the "three Vs" of Big Data—volume, variety, and velocity:

- Volume has to do with scale, as indicated by the IDC's statistics above.

- Variety has to do with a number of new informational formats, as IBM notes in their report *Analytics: The Real-World Use of Big Data*: "The digitization of virtually 'everything' now creates new types of large and real-time data across a broad range of industries. Much of this is nonstandard data: for example, streaming, geospatial, or sensor-generated data that does not fit neatly into traditional, structured, relational warehouses."[6]

- Velocity has to do with the ever-increasing speed at which all of these data are flowing into servers or the cloud where it's stored until ready for use by an organization. Typically as a first way to analyze huge pools of data, organizations will use a platform like Hadoop that allows for distributed processing of data across multiple computers or individuals. A rough analogy for this would be having a massive amount of data placed within Excel spreadsheets to allow for information to be captured without being analyzed.

If you haven't tried to categorize large sets of data, I can tell you from experience that it can be overwhelming to try to make sense of multiple types of information all related to your business or life. But once a framework for understanding has been established, the value of the insights mined from Big Data can be transformative. This is why finding someone to interpret what your data means is of paramount importance.

When the New Oil Is Crude

The phrase "Data is the new oil" is attributed to a woman named Ann Winblad, senior partner at Hummer Winblad Venture Partners. It squarely positions data in an economic sense via a metaphorical comparison to the oil industry. And in the same way that the control of oil can dictate economic advantage, the interpretation of data can steer decisions that can deeply affect the outcome of an individual, company, or organization. Like oil, data also has to be refined for it to have value to an organization, or it remains "crude" and unusable.

The critical decisions of refinement are being left more and more to data scientists, the Merlin-like programmers who work to determine how best to manage an organization's information. Up to now, this role has typically been filled by an IT staffer tasked with organizing company infrastructure to benefit corporate information. But the vast wealth of data at modern organizations requires a new set of skills that includes marketing savvy along with digital expertise. As Claire Cain Miller reports in the *New York Times*, Big Data expertise is so new that curricula or programs haven't been developed for the field. Data science is also such a broad field of study that it's more of an academic discipline than just a smattering of specific courses.[7]

The danger of allowing data scientists to be the sole evaluators of how data should be interpreted is that they may not always know the full context of how it's going to be used. In the Big Data industry, this tension is referred to as domain versus data science expertise. As an example, a marketing-focused employee at an organization may want to understand what the social sentiment about a certain product means regarding a recent sale. However, not being versed in the tools of sentiment analysis, the marketer won't know the specific type of report to ask for from a data scientist.

I discussed this issue in an article for Mashable, "Big Data's Value Lies in Self-Regulation,"[8] with Jake Porway. Jake is the founder of DataKind, an organization that brings together leading data scientists and high-impact social organizations through a comprehensive, collaborative approach that leads to positive action through data. Here's an excerpt from the piece with regard to finding a balance between knowing what to ask for from data and how to interpret it:

"My biggest fear is that data science is used as a blunt tool and that people don't understand the cultural implications of quantifying our world," says Jake Porway . . . [who is] as much a data philosopher as scientist. Gifted at navigating the channel between hacking and hypothesizing, he is adamant about helping people understand how to create context as well as code for insights based on big data. [Regarding this balance,] he says, "This due diligence should be embedded in our craft."[9]

The amount of information available via Big Data means organizations need to break down silos between IT and marketing, whatever people's titles. More data is a blessing as long as everyone works together to make the best sense of what it means.

From Big to Little Data—Profit from the Personal

The term "Little Data" has evolved as a label for the digital information directly relating to consumers. This could be output from your quantified self app or data from sensors in your car relating to your driving behavior. But whereas Big Data refers to a multifaceted organization, Little Data is focused squarely on you. Mark Bonchek describes the evolving relationship between Big and Little Data in his article "Little Data Makes Big Data More Powerful," citing the need for brands to begin empowering customers by helping provide them with this granular information about themselves to make more informed purchasing decisions.[10] Note that "Little Data" in this context doesn't mean there's less information available for analysis; it means the analysis is focused on just your data. The applications for this level of data scrutiny are endless. Patrick Tucker even showed how two researchers could predict a person's approximate location up to eighty weeks into the future at an accuracy level above 80 percent.[11]

If Little Data can predict your behavior almost two years into the future, it should be obvious by now how important it is to get a handle on your data.

In my Mashable article "Big Data's Value Lies in Self-Regulation,"[12] I interviewed Martin Blinder, founder of Tictrac, a platform that aggregates apps to help create and manage user's projects to manage their day-to-day lives. The service is gaining a lot of traction with users, largely because it appeals to the notion of Little Data by functioning as a sort of iTunes marketplace for all of the "life project" apps currently in the marketplace. If users already have a number of different devices, they can aggregate them with Tictrac and focus on a specific type of program, like health or fitness. The company has brokered deals with multiple partners to help achieve this goal, and its platform also syncs with over forty application programming interfaces. Functioning as a data dashboard, Tictrac will eventually begin to learn a user's behavior to personalize information. Blinder believes this type of evolved adaptation will help people design their lives in the future. "By creating a context around lifestyle design, we're enabling a world where anybody can find data relevant to them. I feel we're reaching a point in history where we can really empower ourselves based on an understanding of our own data."[13]

Like many aspects of technology, its greatest power is the ability to disappear so users can focus on improving their lives.

From Broken to Broker

Your personal data gets around. Sometimes it's little, focused on your individual actions. Sometimes it's part of a big picture, a part of an economy that's still in its infancy. Right now, advertisers and data brokers are Hacking Your H(app)iness, analyzing the behavior on what brings you meaning to mine insights for their commercial gain. Once you realize your data is an asset, however, you'll claim it as your own and get rid of the middleman.

You'll become the data broker. You're the agent for your digital identity, the manager for your connected content. If you get upset at the idea of identity theft, you should be livid at the notion that someone else will make money off your personal data. And if you still don't believe me, wait until you see how things will look in the near future.

AUGMENTED REALITY

There is no other future of computing other than [Augmented Reality] which can display information from the real world and control objects with your fingers . . . it's the keyboard and mouse of the future.[1]

MERON GRIBETZ, FOUNDER AND CEO, META

I WROTE A SHORT story in 2012 that describes how I see augmented reality working in our very near future, a (Google) Glass half-empty/half-full scenario to whet your appetite regarding the possibilities of how geeky tech will influence our lives.

SELF-SCREENING

I lurched from the train car, elbow to elbow with a thousand other commuters stepping off New Jersey Transit. I jerked my head to the right and heard a chime indicating my CPRS was online. A bright red arrow hovered in the air before me, analyzing the platform leading to the stairs going up to the main platform of Penn Station.

"Go right." Sean Connery's brogue sounded in my brain as a red line appeared on top of the horde of pressing flesh, all vying for the same staircase. As I turned my head, the line flashed green when my best virtual path appeared before me.

IBM's CPRS (Consumer Pattern Recognition Simulator) lets you set the voice that navigates your actions through a virtual commuter game. (Connery's voice had been chosen for me because I was a fanboy.) The app worked for any major New York transportation hub and was the latest in IBM's Smarter Cities offerings. It utilized image-recognition-based augmented reality to analyze results of multiple predictive formulas to create algorithms based on commuter behavior. The game played out on my iPhone 8 contact lenses.

"What arrr yoo prepared to dooo?" Nice. Connery's quote from *The Untouchables*.

I headed toward the stairs. In my urgency, I bumped a woman next to me, and she grunted. In the upper-right-hand corner of my vision, I saw my points decrease on a small New Jersey Transit con.

"Fuck!" I muttered, apologizing and letting her pass. In my ear I heard the sound of a baby crying and my points dropped even further. The AR in my contact lenses analyzed her past fifty tweets and discovered she was pregnant. Son of a bitch.

Everyone's actions in the game were tied to real-world penalties and rewards. Early social-based action apps like Recyclebank and DailyFeats were still in use to encourage people to earn free stuff or gain social cred. But apps like GymPact where, by choice, you were penalized by your peers for not going to the gym had become wildly popular. Geek-chic went from craving Klout to demonstrating your accountability, and the craze had caught on with local government and utilities companies. I regularly did my laundry at three in the morning to get a high ABI (accountability-based influence) score from OPower, the leading social network based on the Smart Grid.

In my case, my next month's commuter pass would cost about fifty cents more because of bumping a pregnant lady. So now I had to make up my points via speed. I started walking fast. A heart-shaped icon appeared in the upper left-hand corner of my vision as the pulse monitor watch grew snug on my wrist. If the heart went from red to purple, my doctor would get a text indicating I was at risk for cardiac arrest. The monitor went all the way to magenta four times one month and my insurance premiums increased.

Once at Penn Station I headed for the stairs, noting the commuter-gamers outside Starbie's. (Certain sims let you order your coffee mid-play so you could pick it up right away and mobile-pay via NFC.) Distracted by the aroma of fresh-brewed coffee, I stumbled on something large at my feet. I looked and saw a large sack of grain. A money icon appeared in my vision over the bag, so I pulled my eyes to the left, indicating I would take the points for the grain. Frustrated at having to wait for the points to tally, I kicked the sack, hard. Then I ran upstairs.

As I neared the middle of the station, a red stop sign icon filled my vision. I paused, not sure what was happening. Virtual Air Rights codes had deemed it unlawful for advertisements or any game component to trick someone when using augmented reality–based app functionality.

I heard a new voice in my ear. "Chuck, look up."

This freaked me out, as none of my voice recognition software was programmed to speak unless I spoke first. And most of my sims used eye tracking or Microsoft Kinect to recognize my gestures before I heard external voices in a game.

I looked up as I heard the board clicking, the shifting words forming the following phrase:

All the world's a screen, and we are merely layers.

I stood for a long minute, gazing at the board and not really comprehending what was happening. I vaguely registered that my contact lenses were in reality mode, meaning the board actually said the words I saw above me.

"You a Shakespeare fan?" the voice came again.

"Sure?" I said, turning to see where the voice was coming from.

"Up here, Chuck." I looked back at the board, and one side of the screen was shaped like a smiley-face icon. The lips moved when it spoke. "Thought I'd go with a smiley face versus a scary *Tron*-looking thing. Besides, part two sucked."

"Agreed."

"They call me the Bard. A few years back MoMA did a real-time data exhibit thing and they ran Shakespeare quotes on my screen. Somebody got cute and took the 'o' out of 'Board' and it trended on Twitter, so here we are."

I looked at the other side of the screen, opposite his "face." "So what's with the quote? Am I a 'mere layer'?"

I assumed the metaphor had to do with the Smart Grid, where the notion of Big Data meant that with networked artificial intelligence, we'd arrived at an Internet of Things mentality. In a sense, everything with a chip in it was alive, or in this case, a layer. Being a geek, I'd felt the whole idea of the singularity was inevitable starting around 2011 or so. I was also sure Bard was hooked to the Internet and dozens of cameras in the station that pumped images he could access anytime from the cloud.

"Sort of," Bard responded. "Don't get pissed, but I accessed your e-mail and social channels just now."

I wasn't that pissed. "Privacy" had a whole new definition these days. Since a GPS knew where you were at all times and everyone's virtual games were hooked real-time to the Web, government types simply accessed video or Outernet feeds from citizens any time they wanted. Everyone's lives were recorded at all times. According to CNN, no event occurred without at least two cameras recording what happened for potential public usage. News and law enforcement had become whole different animals in the past few years.

"Why is my foot wet?" I said, interrupting Bard as warm liquid seeped onto my right foot. I looked down to see my black loafer had a dark stain.

Bard spoke quietly. "That's what I'm trying to tell you, Chuck. Look."

His screen switched to an image of me arriving at the platform downstairs from a few minutes ago. I saw my journey from the train and could tell I was in my game, since my eyes appeared glazed and distant. The camera views switched a few times, from commuters to station cameras and back. And then as I turned a corner, I saw myself stumble where I had kicked the sack of grain.

But it wasn't a sack of grain. It was a homeless guy.

He had blocked my path, and I had stumbled hard into his arm. And then I watched in horror as I pulled my foot back and kicked him squarely in the face, breaking his nose. Blood poured onto my shoe as he clutched his face in agony while I simply took a step back while my points tallied.

"You weren't supposed to kick the grain, Chuck," said Bard. "That's what the money signs are for. The game disguised Tom as a bag of grain so people would avoid him. Kicking good things means you lose points. But you were too fast. And Tom sat up at a bad time."

I couldn't move. That image of me kicking the homeless guy—Tom—kept playing over in my brain. It was an image I knew I'd never erase. And it wasn't a game.

What kind of man am I?

I stood for a long moment, game-blinded commuters rushing by. For once I heard the sound of shoes on pavement—no sound track, no sound effects. This was reality. And it sucked.

"Chuck."

I looked up. Bard had cleared his screen and the following words appeared slowly, one by one:

What are you prepared to do?

About a year later, on my birthday, Bard said he had a surprise for me.

"Turn on your CPRS game."

I did, and my commuter sim turned on. Bard had hacked it so it was pointing downstairs, and I followed the arrow to the spot where I had kicked Tom. Instead of a sack of grain, I saw a huge virtual package with the words *The Impossible Idea* written on the side. An icon on the upper-left-hand corner of my vision flashed, indicating it would open if I moved my eyes quickly to the left.

My impossible idea had been fairly simple. I quit my job and volunteered at the Robin Hood Foundation to create an app that rewarded people for kind actions to the homeless in New York City. Gwyneth Paltrow was their spokesperson, and with her avatar in the sim, the game took off. Starbucks joined in, and pretty soon the pay-it-forward mantra went full swing in Penn Station. Within a few months, people donated their Klout perks and accountability bonuses so that actions generated behavior change as well as words.

So now I looked at the spot where I had kicked Tom and my life had transformed by mistake. I moved my eyes to the left, and the virtual package fell open.

And I saw Tom. He was clean-shaven, waving and smiling. He was piped in via Skype and virtual, wearing the Robin Hood T-shirt they'd given him the day we first met. He'd gone from being a volunteer to a full-time staff member and gotten the first assisted-living residence made via profits from my app.

I smiled. "Thanks, Bard."

Staring at the spot, I blinked three times rapidly to turn off my sim contact lenses. My vision cleared of all icons, layers disappearing between me and the empty pavement where Tom used to lie.

And it was empty.[2]

From (Cyber) Punk to Possible

I've been a science fiction fan since I was a boy, watching *Star Trek* with my dad and brother, and then seeing *Star Wars* in the theater when it first came out. Years later I devoured the work of people like Arthur C. Clarke and Robert Heinlein before falling into a geeky bromance with cyberpunk fiction from writers like Philip K. Dick and Neal Stephenson. Dick wrote *Minority Report*, which was made into one of the greatest cyberpunk films of all time, featuring Tom Cruise as a police officer who utilized augmented reality technology to stop murders that hadn't yet occurred. Augmented reality, or AR, is a term coined by Tom Caudell from Boeing in 1990, as noted by Brian X. Chen in *Wired.*[3] Caudell created the term based on a head-mounted, hands-free interface used by workers assembling airplanes. Digital data overlaid on the glasses in front of workers' eyes let them see reality (the plane) with augmentation (instructions in real time on what to fix). Beyond the convenience for workers, the technology set the stage for combining virtual and physical reality.

There are a lot of desktop-based applications of augmented

reality utilizing your webcam, but my fascination has always been with mobile AR. Whether you're using your phone screen, a tablet, Google Glass, or eventually contact lenses, the idea of hands-free mobile computing is transformational. When images appear in front of your vision based on what you're looking at, the technology simply feels magical.

I first wrote about AR for *iMedia Connection* in the article "Augmented Reality: What Marketers Need to Know." Here's how I described the future of AR:

> The seminal promise of AR is as the touchstone technology allowing social networks, geo-based tracking, and the semantic Web to converge. Put less geekily, think of AR as your personalized digital butler, who will get to know your behavior so specifically that it can prethink your choices based on your friends, location, and how you search online. The cyberpunk fictions have come to reality with AR, and the cultural ramifications are as powerful as the marketing opportunities.[4]

At the time I wrote the piece, most of my marketing friends felt AR was gimmicky and only useful for gaming applications. But I saw AR as a type of wearable computer, an idea that's pretty simple—you take the mobile phone you're already carrying (which is a computer) and put it in front of your eyes to access information. The big differentiator for augmented reality, however, is how digital data can be placed over what you're seeing on your mobile screen based on what you're looking at. In this way, as I often say, augmented reality's main staying power comes in the fact that it's a *browser*[5] versus just an application. It lets you look at digital data you can't see unless you're using AR technology.

I call this greater world you look at via augmented reality the Outernet. Instead of having to turn on your computer or look down at your phone, when you're using an AR-based technology, the

Outernet is already all around you. You're used to this idea, for instance, if you've got an Xbox and use the Kinect gaming system. I wrote about this in *iMedia*: Kinect recognizes your movements using AR with video games as you play in your living room. Using something called haptic technology, you don't even need a controller; using your hands and body, the system recognizes your movements that characters in the game respond to.

There's another great AR game created by Georgia Tech called ARhrrrr that lets you kill zombies by using your mobile phone and their augmented reality app. By pointing your phone at a physical map that serves as an image marker for the game, the system recognizes your movements as you move around and try to kill zombies. An intriguing feature of the game also comes when you place a physical object on the map, like a Skittles candy. The creatures in the game react to the candy that acts like a bomb in the augmented environment. A consumer application of this technology would be for a coaster at a restaurant: Kids could play an AR game while waiting for a soda.

Technology oftentimes gets adopted faster for the pragmatic uses it offers versus a more gimmicky application. In that sense, one of my favorite augmented reality apps is New York Nearest Subway from a company called acrossair. I've lived in the greater New York City area for over twenty years, and I still get lost in SoHo. When you look through your mobile phone screen using New York Nearest Subway, you see a visual icon based on the nearest subway transit station. So if a floating letter "N" is in the right-hand part of your screen, turn to the right and head to that train. It's a great way to get a sense of how pervasive AR will become in the future with these types of applications.

The possibilities for AR technologies are limitless in the same way that Internet applications know no bounds. The primary reason the technology hasn't become ubiquitous yet is that it's only available on smartphones, and things like phone processor speeds have been slow up until the past few years. But those issues have been diminishing as time wears on. There was also an era when physical markers were needed for augmented reality to work (black X-shaped boxes similar to QR codes). Then image recognition came into play, where you can simply hold your phone up to a photo or object that is recognized by the AR tech and projects a digital image over your screen.

Here are some other examples of how augmented reality is currently being used:

- In cars—General Motors created an AR-enabled windshield.[6]
- In surgery—the Scopis Surgical Navigation System provides guided visualizations.[7]
- In retail—German AR company Metaio provided Ikea with an AR-enabled catalog.[8]
- In dating—the short film *Sight* shows AR and facial recognition used for dating.[9]
- In art—projection mapping on walls uses AR light shows to entertain.[10]

The Future and the Financials

"We believe that physical real estate will become a valuable commodity once augmented reality–capable devices are ubiquitous." Wedge Martin is a cofounder of GeoPapyrus, an AR company that lets you publish and interact with social content such as photos, videos, audio, or websites by browsing physical elements (frames, windows, buildings, books) from the real environment around you. I asked Wedge why he created the company.

Right now, people pay ten thousand dollars a month to advertise on a billboard when they have no idea how many people will look at it. We want to find ways to increase engagement for these types of environments as well as other physical spaces. In essence, if we increase people's interaction with the physical side of things, we'll be bringing "social networking" back into the real world—that place where people walk around and get fresh air we feel has been hugely underrated as of late.[11]

Wedge and GeoPapyrus are addressing an idea I wrote about in Mashable a few years back called Virtual Air Rights (VAR). Here's how I explained this concept in the article "Who Owns the Advertising Space in an Augmented Reality World?":

Look up in Times Square and you'll see the earliest version of a banner ad. Real estate developers pay massive sums to secure air rights for the empty space above buildings. Monetizing by building up (as opposed to out) in crowded areas like Manhattan, they also get to dictate what advertisements appear in the air that they control. Augmented reality (AR) has made it possible for this same paradigm of advertising to exist via your smartphone. Multiple apps feature the ability for ads to appear on your mobile screen as miniature virtual billboards assigned to GPS coordinates.[12]

Virtual Air Rights will be a fascinating subject in the coming years. As an opt-in experience like the one GeoPapyrus provides, VARs provide a huge economic opportunity. In essence, the Outernet is a blank canvas ready to paint, and advertisements will come to look more like experiences or content than billboards.

Along with opportunity, there's going to be controversy. After the BP Deepwater Horizon Oil Spill, Mark Skwarek and Joseph Hocking created an app called The Leak in Your Home Town. Just point your phone at the BP logo after downloading the app and it bursts into flame. It's an amazing bit of geekish parody, but an area of law we'll hear more about[13] in the years to come.

For instance, in the near future you'll walk into a grocery store wearing an augmented reality–enabled device and only see the brands you want to buy. You'll input your grocery list and, if you're a Pepsi fan, you'll only see their products in the soda aisle. Items from Coke will essentially disappear, appearing as an empty shelf if your device looks in their direction. The physical real estate at the ends of aisles in the grocery store known as endcaps will also be highly desirable *virtual* real estate.

Our faces are also billboards, and there's the possibility that people will screen other faces they don't want to look at utilizing AR. Irritated by Democrats? Using facial recognition, you'll only see images of a donkey over people's faces if they vote Democrat. Taken to the extreme, this could even lead to a form of virtual racism if you don't want to see people of a certain race or background.

In terms of facial recognition and AR, I asked Senator Al Franken his thoughts on the issue, per his role as chairman of the Senate's Judiciary Committee on Privacy, Technology, and the Law:

Facial recognition technology is a big deal. It can help us catch dangerous criminals and secure sensitive workplaces. But it also places a tremendous amount of power in the hands of governments, companies, and private individuals. The technology already exists that will allow a stranger to identify you, by name, by simply snapping a photo of you on the street. That's a problem. At a bare minimum, facial recognition should only be deployed commercially—and on a strictly opt-in basis.

What's worse is that our privacy laws are utterly unprepared for this technology. Over the past few years, Facebook has used the photos posted to its site to create the world's largest privately held database of faceprints—without people's permission. If you have a Facebook account and you haven't clicked a little button on a little menu, chances are that Facebook has created a unique digital model of your face. That file can be used to identify you in any photo taken anywhere, whether or not that's posted to Facebook. All of this is 100 percent legal under federal law.[14]

Rather than try to hinder innovation, however, Franken's primary focus is on helping consumers be informed about new technologies. He also urges companies to be proactive in having consumers choose to opt *into* a service, rather than having them automatically be signed up for a service without their knowledge where they have to opt *out*.[15] Regarding Google and Glass for this issue, Franken noted:

In the past, Google has taken a thoughtful approach to facial recognition technology. While facial recognition is on by default on Facebook, it's opt-in on Google+. I get the impression that the Glass team is also taking the privacy concerns around facial recognition quite seriously. My biggest goal is to make sure that our privacy laws keep up with our technology. I want to make sure that all of the benefits that we see from new technology don't come at the expense of our privacy and personal freedom.[16]

The legality around issues of augmented reality are rapidly developing. Brian D. Wassom is the Social, Mobile, and Emerging Media Practice Group chair and a partner at Honigman Miller Schwartz and Cohn LLP. He has an excellent blog on AR and legal issues, and recently wrote about best practices for facial recognition privacy.[17] I interviewed him for *Hacking H(app)iness* and asked him if people would be able to block their images in the future from being recorded by Glass or other such devices.

The actual act of capturing someone's image doesn't necessarily infringe their publicity rights. Indeed, the creation,
reproduction, and distribution of imagery is the subject
matter of copyright law, and typically the person who creates the image owns it. What you do with the image—i.e.,
whether and how you commercially exploit it—determines
whether you're infringing on publicity rights.

That said, I fully expect publicity rights law to evolve in
response to situations like this. Precisely because we don't
have a uniform rule describing what this right even is, let
alone what it protects or how it can be infringed, it leaves a
lot of room for creative interpretation by opportunistic law-
yers. For example, although I have yet to see anyone argue
this, I wouldn't be surprised if we soon hear the argument
that individuals' facial features are a part of their identity,
and that they can be exploited in various commercial ways,
and so any use of unauthorized facial recognition technology is a violation of that person's publicity rights. Mark my
words—that argument is coming.

Of course, publicity rights and copyrights aren't the only
legal issues implicated by surreptitious recording. The scenarios you describe raises questions of privacy and eaves-
dropping, first and foremost, which also vary by state.
Applying these laws will require fact-specific analysis to
determine whether there was an expectation of privacy in
the specific circumstances involved.[18]

A number of companies are helping navigate consumers
toward a time when augmented reality technology becomes ubiquitous in a privacy-protected environment. One of the best applications for AR in this regard is in the B2B, or business-to-business,
environment. A leading company in this field is APX Labs, which
is focused on creating AR environments for "deskless workers"
using their smart glasses technology. I interviewed Robert Gor-

don, the company's chief strategy officer, to ask him more about this trend.

> Deskless workers, enabled by smart glasses (augmented reality–enabled glasses), provide employees with a hands-free environment so they can be more efficient. Doctors and nurses in the medical community, workers in transportation and logistics, even people in the entertainment community are better off when utilizing hands-free, wearable computing. Biometrics can also be a component of the smart glasses experience.[19]

Biometric integration provides a fascinating use of technology that APX has incorporated in the past with military technology. Outfitted with smart glasses that can utilize a technology like Cardiio to read people's heart rates by measuring changes in the skin tones of their faces, military guards can determine if someone they're speaking to is getting nervous.

A medical application of this same technology could allow a doctor to look around a waiting room with smart glasses and be alerted to the patient who most quickly needs attention, based on heart rate or other visual characteristics. APX has also created something called See What I See (SWIS) that lets people wearing smart glasses switch views to what another person is looking at. For health workers in the field, this means they could stare at a sick child and get a consultation from an expert physician anywhere around the world. You can see how this experience looks in a consumer environment by watching a video created using Google Glass by a teacher taking his class on a virtual field trip.[20]

The Vision's Our Own

Technology isn't inherently good or evil. People can use a device or platform as they see fit. Augmented reality used for value-added applications is already transforming the world for good, allowing people to easily see information they haven't seen before in a hands-free environment. Facial recognition technology can also be utilized to great benefit when people knowingly allow their images to be tagged or utilized for medical or work purposes based on mutual consent. Soon biometric and quantified self data will also let people project a visualization of their own health to the world, perhaps publicly showing something like an icon of a sneaker to be identified as a runner in public settings.

Hacking H(app)iness will also mean we can project our emotions to the world, where augmented reality applications could essentially let our faces or clothes be perceived as a sort of mood ring to others. Or our faces could serve as the permissions portal for economic exchange, where selling your data to trusted sources could provide a new source of income. The visions of the future are limitless, and with augmented reality, we'll get to see them like we never have before.

VIRTUAL CURRENCY

All the artifacts of a human being belong to physical and logical governments, and not to social networks. But the ability to move any form of asset between the virtual world and the physical world needs a commonality of understanding of identity.[1]

J. P. RANGASWAMI

SOCIAL CURRENCY is an idea you're already used to. In a particular clique of friends or at your local church or synagogue, you've earned a reputation. Perhaps it's based on your words and personality—you're outgoing and funny. Or perhaps your reputation is based on action—you volunteer to help a lot. Wherever you fall into that spectrum, you've earned a social currency based on your identity that is made up of a combination of your words and deeds.

You're also aware of how you can redeem currency in these types of real-life social networks. For instance, say you helped someone from your synagogue carry a heavy box from inside the building to their car after a service. If you asked them to help you move a box of the same size the following week, you'd expect them to help you. If you asked them to help you move out of your *house*, however, you'd be straining that relationship because it's not perceived as a fair exchange of value.

A big part of Hacking H(app)iness is understanding how the worlds of digital technology, economics, and happiness intersect. Economics in the digital realm is something we've already touched on—you may earn a high enough Klout score, for instance, to get a Klout Perk. A brand offers you something based on your influence and the value exchange is understood. You get a coupon or a product of some kind, and the brand gets the benefit of the marketing you generate when you talk about your perk to your Twitter or Facebook audience. This is essentially paid advertising, where the brand reaches your audience in exchange for a product.

Economics and happiness in the digital realm intersect with regard to your data. As consumers and citizens we're being measured in ways we don't fully understand, but the portraits of who we are (our digital/virtual identities) are being projected in public and private realms. If we don't understand or control our identities, how can we measure or even increase happiness? To fully leverage information about our happiness and well-being generated from quantified self devices or other inputs, we need to own our data.

But what if you could sell your data directly to brands at a profit? By eliminating data brokers, people could establish their own virtual storefronts for their own data to sell as they'd like. This would provide a form of virtual currency exchange, as the online market would determine what your data was worth.

Federico Zannier, a New York University graduate student, decided to pursue this idea by selling his data via a Kickstarter campaign called A Bite of Me. People who participated in his campaign bought increments of Zennier's data beginning at two dollars a day. Zannier utilized a number of tools to gather data for the project while also educating users. Like marketers, Zannier used tools that tracked his Web traffic and activities, and software that tracked his GPS location. But as Sarah Kessler explains in her *Fast Company* article "What If We Thought More Often About Being

Tracked Online? Man Stalks Himself to Find Out," it was the act of selling his data via Kickstarter that made Zannier's project truly unique. First of all, it made people realize how thoroughly our actions are being measured and evaluated via our online and mobile behavior. Second, as Kessler notes, the project poses a question: If data is valuable to companies, why shouldn't the people who create that data be able to sell it?[2]

I bought a week's worth of his data, paying five dollars, which included around five hundred websites Zannier visited, four thousand screenshots he took, four thousand webcam images, a recording of his mouse pointer movements, his GPS location, and an application log of 4,200 lines of text. Here are some visual samples of his data:

Do I know what an advertiser would pay for his data for this same time frame? Nope. I am fairly sure if I did the same thing, an advertiser would pay a lot less for *my* data, as I'm forty-four years old and have two kids and a mortgage. You don't need a calculator to figure out that means I have less disposable income than a young, savvy New York graduate student.

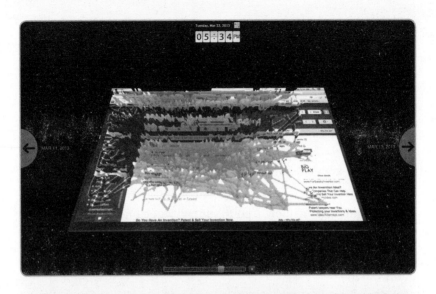

The more relevant question is: Do I feel a week's data is worth five dollars? Apparently so, because I bought it. For his effort alone the money was well spent, and I'm also paying to support his educational message. Plus the act of paying him made me happy. I genuinely got pleasure knowing that I was supporting his work versus giving money to a third-party data broker neither of us will ever meet.

I interviewed Eli Pariser, author of *The Filter Bubble: How the New Personalized Web Is Changing What We Read and How We Think*, about this issue. While he wasn't sure that selling personal data will work at scale, he does see the value in trying.

> It's hard for me to envision a world where people could make a fair amount of money selling their personal data. However, I would love to move toward a world where people are able to leverage their data at least as well as big companies do. Apple knows more about my behavior from my iPhone than me.[3]

My transaction with Zannier demonstrates a form of virtual currency, one where I helped him leverage his data for fiscal gain. I trusted that his data was worth five dollars. And I won't be upset if I find out it wasn't worth five dollars. Our exchange was built on trust, even though I've never met him. Trust ranks high in importance for the happiness economy. That's why your actions, as reflected in data, are such a big deal. Other people's algorithms, like the ones being used for a dating service, will judge whether or not you're worthy to meet based on your reputation. Learning about virtual currency will help you understand where the banks of the future will be tallying information relevant to you and your relationships.

Virtual Currency, Past and Present

I wrote about one of the most famous examples of virtual currency for my Mashable article "Why Social Accountability Will Be the New Currency of the Web":

> In 1932, a small town in Austria called Worgl created an economic experiment to counter the devastating effects of the

Great Depression. The mayor issued a new currency and encouraged citizens to spend it quickly to put money back in the system. People were motivated to participate in an economy based on action. Within months, the town's unemployment rate had dropped by over 30 percent. Dubbed "the Miracle of Worgl," the experiment was eventually terminated by Austria's Central Bank in 1933 for fear the nation's existing currency would lose relevance. Unemployment immediately returned, and Austria's economy collapsed further in the wake of Hitler's rise to power.[4]

One wonders how Austria might have fared differently if the Worgl idea had spread through the country before Hitler came to power. Perhaps greater economic strength would have helped them avert the Anschluss.

This demonstrates how value exchange by consenting parties could work to everyone's advantage. It is similar to an economic model known as "the commons," in which citizens work together without needing oversight from government or other institutions to advance society.

Elinor Ostrom, the most famous proponent of the commons and the only woman to ever win the Nobel Memorial Prize in Economic Science (while not actually being an economist), demonstrated with her research that a number of communities flourished by exchanging virtual currencies rooted in social capital. Neighbors essentially decided among themselves how to pool and distribute resources, and monitors were set up to fine or eventually exclude anyone who broke communal rules.[5] What's apparent from these aspects of the commons is the inherent trust that pervaded the communities who set up these guidelines. Where rules were adhered to by individuals, positive reputation and trust in the communities increased. In our lives today, online etiquette has created a form of digital commons. There's a reason the media is called *social* after all—how we relate to other people determines our currency or value.

Social Networks as Banks

It may surprise you to learn that some social networks operate as banks, as they're the primary platforms for certain types of monetary exchanges beyond traditional online purchases. For millions of users, Facebook Credits operated for a number of years as the primary form of currency when purchasing in-game items, for example (the company stopped using Credits in 2013). One U.S. dollar equaled ten Facebook credits. In 2011, Facebook began requiring that game developers process payments only through their Credits platform. Facebook kept 30 percent of all revenue collected, and developers retained 70 percent. While Facebook's gaming revenue declined in 2012, the company's third-quarter profits that year were still a sizable $176 million (U.S.).[6]

Thirty percent is a hefty fee for developers to have to pay to Facebook. While the network is hugely popular, developers still have to pay to drive people to their Facebook pages while trying to make revenue from their games. Fair or not, the example points out how Facebook was acting as a financial institution by taking direct payments from gamers and charging a fee for developers while mandating they only take payments through their Credits platform.

The trend for digital and social networks to act as banks is growing, as reported by a press release from Gartner in October 2012. Noting the advantages networks have over banks, the release noted:

> Digital mega-firms have many things in their favor. They are masters of data management and analytics. To all intents and purposes they define agility, both from a technology

and a business model point of view. They are extremely adept at extending their value chain analysis beyond the core offering, with an eye to identifying new opportunities for business and highlighting specific customer needs that they might address. They have the ability to define—and then redefine—the business models that they deploy while their focus on what partners can bring to their propositions stands as an equally strong differentiator.[7]

There's a concept in telecommunications known as Metcalfe's Law. The technical definition is that the value of a network is proportional to the square of the number of connected users of a system. In plain terms, this means you need two devices speaking to each other to have any value—for example, owning a fax machine is useless if nobody else has one. The concept has now spread to users. You've already experienced Metcalfe's Law in your own life, the first time you saw someone with a device you didn't have. If you're my age, you saw someone with a cell phone the size of a shoe box back in the day and mocked them for six months until you had to have one of your own. Perhaps you held out longer than I did, but by and large you got tired of people asking, "Why can't I reach you on your cell?" So you caved and bought one because everyone else was benefiting from a device you didn't have.

Virtual currency has this effect if you've ever wanted to make a purchase and had to use PayPal. You may want to pay with a credit card, but if the only way you'll get a product is with PayPal and you really want it, you signed up and got an account. Metcalfe's Law will also apply to things like Google Glass. While a recent study cited that only 10 percent of Americans say they would wear Google Glass, it didn't take into account how the other 90 percent would react when seeing people wearing the device (outside of the fact that 10 percent of Americans adopting any new product is a staggeringly large number).[8] By the time augmented reality glasses from any brand become cheap enough for the average consumer to buy, the early adopters will influence their friends who will in turn purchase devices, and we'll see the Outernet get turned on at scale.

In May 2013, Amazon created its own form of virtual currency by giving a set of Amazon Coins to Kindle Fire users to buy apps and games.[9] Like Facebook Credits, the plan was also geared to try to increase developers' interest by making more money off apps when customers are incentivized to purchase more with the Coins. In response to this announcement, an article in the *Economist*[10] summarized a key point of David Graeber's book *Debt: The First 5,000 Years*, describing how the idea of cash money (versus trade) first developed. To help soldiers get food and provisions, a king created coins his subjects had to use to pay taxes. Then he gave these coins to his soldiers to buy provisions, while subjects got the coin of the realm needed to stay out of prison. Fast-forward a few thousand years and the origin of this process has been lost, but the metal or paper currency still holds sway over our collective imaginations when it comes to value creation.

It's easy to dismiss virtual currency, thinking it's only part of a social network so it's not really money. But Graeber's point about the army shows how cash money first came into being—by the leader of a sovereign network (or nation) declaring that a certain form of currency should be utilized. It's the same process that happened in Worgl and Facebook, just enforced for a longer period of time.

Just because something is digital doesn't mean it's not worth dollars.

In my interview with J. P. Rangaswami of Salesforce.com about virtual currency, he pointed out another example of digital dollars many people didn't consider: "[In 2011] Steve Jobs said something in relation to Apple most people didn't report on. He pointed out that Apple had over two hundred million active iTunes accounts that were associated with e-mail addresses and credit cards—meaning Apple has the most accounts with cards anywhere on the Internet."[11]

While purchasing movies or music from the iTunes Store is done with credit cards, Rangaswami's point focuses on the same idea that Graeber noted about a sovereign creating currency: Once you're part of any closed network, the people running the network control its economy. You may not think of a sale within iTunes as a form of virtual currency, but it is. Whether it's positioned as an actual coupon or an incentive to buy, it's simply a form of digital haggling; you *can* buy just one episode, but if you buy all ten you get a discount.

Programs offered by airlines to get free miles operate in the same way. You're incentivized to make purchases to gain more travel. You're agreeing to an exchange because you see a fair value for both sides. These programs are a type of virtual currency you've been used to for years.

From the Virtual to the Virtuous

It's important to recognize that you're used to virtual currency when you think about Hacking H(app)iness in the personal data economy. If your data is protected in a data vault, in essence you have become your own economy. In the same way Federico Zannier sold his data, you could, too. People may not buy it, but the point is you're controlling the currency in your own economy. While your data may be virtual in one sense, the currency or money you'd make from selling it could be used to buy actual goods.

Now add to the idea that you're your own economy and think about trust. If you've got a positive reputation from a certain community, online or off, this will help get you get more money when you want to make a transaction. For one thing, people probably like you if they trust you, so they want to help you out. They also feel you'll probably help them in the future. So now the main issue you'll need to solve is what you can sell that people will want to buy.

What if people paid you for kind deeds? When virtual currency takes off, this practice might be called "life-tipping." Instead of tipping in a physical sense, like throwing money in a hat after hearing a band play, you'd assign someone virtual currency when they've done something positive. When augmented reality becomes pervasive, we'll all be able to see digital data associated with someone's identity. This means we'll be able to tag other people in some way, a sort of virtual instant message framework where we can exchange digital information. One of the things we'll exchange is payment. Right now, you can get a Starbucks card that pays rewards when you use it at any location. Life-tipping around coffee is an easy scenario to envision. Someone lets you cut them in line, you nod your head using the Starbucks app for Glass, and the person behind you gets their coffee for free.

Other examples of life-tipping with virtual currency could involve reducing guilt. You may feel remorse when you don't recycle all your plastic bottles. How about life-tipping the homeless woman making three dollars a day pushing a broken shopping cart, recycling bottles from the trash? You don't need to look the other way when she smiles at you now—she's providing a service you don't have time to do. Smile back and life-tip her.

At work, maybe you know someone who is struggling to afford college tuition for their kids. But their disposition is always positive, and any meeting they attend involves laughter. If you managed this person, you could life-tip them tuition vouchers or another form of payment. This form of life-tipping would also eventually be a write-off for your taxes.

Community life-tipping would take this idea to scale and already has a precedent in a "Peace Corps for Caregivers" model as reported on in the *New York Times* article "A Volunteer Army of Caregivers."[12] The article describes the efforts of Janice Lynch

Schuster, a senior writer for the Altarum Institute, who created a petition for We the People, a site created by the Obama administration. If someone gets at least 150 signatures, they can post a petition that will be acted on if it reaches one hundred thousand signatures. Her idea for the petition: "Create a Caregiver Corps that would include debt forgiveness for college graduates to care for our elders." The idea represents a simple value exchange— older adults get companionship and help with basic needs while high school students get tuition credits for college or recent grads get debt forgiveness. While the petition didn't get the hundred thousand required signatures for action to be taken by the government, it did get a large amount of press that pointed out the significant burden boomers will have on our economy once they need assisted care. Schuster's petition raises the question: If there will be more jobs than workers available in the future in these situations, why not alleviate student debt and help seniors at the same time? If kids could be properly trained, the virtual currency exchanged would solve a great deal of problems.

These types of scenarios, on a personal and communal level, will become more prevalent as the personal data economy comes into full swing. When people control their own personal economies, they'll get to decide who they want to share value with, in whatever form.

Sample Virtual Currency Models

For a comprehensive introduction to virtual currencies, an excellent resource is a paper created by the European Central Bank called *Virtual Currency Schemes*.[13] Defining a virtual currency as "a type of unregulated, digital money, which is issued and usually controlled by its developers, and used and accepted among the members of a specific virtual community," it provides a detailed explanation of how these currencies relate to existing banks. It also provides a good synopsis for reasons to implement virtual currency schemes.

> By implementing a virtual currency scheme focused on the online world (basically for virtual goods and services) a company can generate additional revenue. The use of virtual currencies can help motivate users by simplifying transactions and by preventing them from having to enter their personal payment details every time they want to make a purchase. It can also help lock users in if, for instance, it is possible to earn virtual money by logging in periodically. If users are asked to fill out a survey or to answer other questions in order to earn extra virtual money, users reveal their preferences, thereby providing valuable information for commercial use.[14]

In terms of how to introduce these ideas of virtual currency, an organization called Innotribe created the concept of something they call the Digital Asset Grid. The video documentary about the project[15] was shown at an event called Sibos, which is an annual conference presented by SWIFT, the global provider of secure financial messaging services (you may be familiar with the term "SWIFT codes" referring to your credit card). Here's a description of the Digital Asset Grid from the Digital Asset Grid Session at Sibos:

> The idea is fairly simple: While many of us share digital assets every day and store them on various websites, these are, for the most part, assets of low value or low consequence. Few of us would feel safe conducting a complex banking transaction on Twitter or Facebook. In contrast, the Digital Asset Grid would provide a way of conducting transactions involving any high-consequence digital asset on the

open Internet but with SWIFT-grade security and privacy. The Digital Asset Grid is conceived as a set of services built on top of the open, standards-based Internet. SWIFT is working on the Digital Asset Grid to position banks as platforms upon which online services can be built.[16]

A video showing examples of how the Digital Asset Grid could work provides the case study of a woman buying a motorcycle. After seeing an ad for a motorcycle she likes from a trusted seller, she asks for some further information (photos, maintenance reports) that are provided via a protected data exchange. Meaning only the data requested is provided between the buyer and seller. She likes what she sees, goes for a test drive, and buys the motorcycle on the spot using the Digital Asset Grid. All associated digital assets of the motorcycle also change ownership, including things like the insurance policy and maintenance history of the vehicle.

Ven is a virtual currency that's growing in popularity. I interviewed Ven's founder, Stan Stalnaker, for my Mashable article "Why Social Accountability Will Be the New Currency of the Web." Stalnaker had a number of fascinating insights about the nature of evolving digital currency.

> "Facebook will become the biggest bank in the world," says Stan Stalnaker, the founding director of Hub Culture, a social network that revolves around a virtual currency called Ven. "This will happen the moment they allow for P2P exchange of Facebook Credit between users. If they can link that to Likes, and map the value of Likes and other activity on their imprint of the social graph, these values will begin to function like money."[17]

But Stalnaker has already created this P2P exchange via Hub Culture where, like citizens of Worgl, members are expected to put

Ven into virtual circulation as much as possible. Based on a portfolio of units that includes leading currencies, commodities, and carbon futures, the Ven is less volatile than other global currencies and is traded for everything from knowledge to travel discounts and even a Nissan Leaf.

Stalnaker recognizes that the notion of virtual currency is in its infancy. When the disparity of definitions surrounding currency and influence someday merge, a singular value will reflect a common exchange of goods. Until then, he notes that "what currency really is . . . is language. We all speak in English dollars, and some people speak in rubles. What the Internet needs is its own language for currency."[18]

I Speak H(app)y

Accountability-based influence provides a language for currency that should be used in the Connected World. Once we build our own individual economies, our actions more than our words will build trust. Once augmented reality becomes pervasive, people will also digitally tag or life-tip us depending on their perception of our actions. Virtual currency will soon complement the paper in our pockets, and someday remove it for good.

SHARED VALUE

Shared value is not social responsibility, philanthropy, or even sustainability, but a new way to
achieve economic success. It is not on the margin of
what companies do but at the center.[1]

MICHAEL E. PORTER AND MARK R. KRAMER

MICHAEL PORTER is the Bishop William Lawrence University Professor at Harvard Business School and the most cited author in
business and economics from the *Harvard Business Review*. Within
corporate circles he is generally recognized as the father of the modern strategy field.

For many people in the business world, this concept of shared
value doesn't compute. Survival depends only on increasing quarterly profits. It's essential to have a good product or service to
increase revenue, but helping others is left to people in public relations or corporate social responsibility. People may feel good about
their companies giving to needy causes, but philanthropy or charity aren't respected with regard to the bottom line.

But shared value isn't about charity. Shared value, as the name
implies, means identifying all the key stakeholders in a value

chain for a business and seeing how everyone can benefit. There doesn't need to be a moral imperative to help others. Identifying how everyone can benefit means you sustain your business longer than competitors who focus only on short-term gains.

A case study of shared value success comes from specialized coffee company Nespresso, one of Nestlé's fastest-growing divisions. The company has enjoyed an annual growth of 30 percent since 2000, largely based on its dedication to working with coffee farmers to grow the brand.[2] Most coffee farmers around the world work in impoverished rural areas and suffer from low productivity or quality of product due to their circumstances. Realizing that the core of their business depended on reliable sourcing, Nespresso made a business decision based on procurement needs: Establish local facilities near farmers to pay premiums for coffee beans while also providing education and equipment needed to produce higher-quality and more sustainable crop yields. The initiative was hugely successful and is now globally recognized as a case study to emulate with regard to maintaining a happy and healthy supply chain.

It's easy to read this case study and think Nespresso was offering a form of philanthropy to its coffee growers by providing training and equipment that was able to increase farmers' salaries. That's not the case. Working with farmers was a core business decision not based on charity. While the company may have taken some short-term losses to purchase new equipment and provide training, long-term profits increased dramatically. Farmers were able to produce a higher-quality product, and that meant Nespresso outsold competitors. Greater levels of production offset salary increases, plus farmer loyalty meant attrition rates dropped. Shared value means higher profits in the long term. Employee loyalty also engenders a higher level of company morale, but it's not a primary goal. Shared value provides a tangible economic benefit to an organization, not just philanthropic goodwill.

I wrote an article for Mashable called "Social Responsibility: It's Not Just for Brands Anymore" where I compared the trend of shared value in the enterprise to accountability-based influence shaping people's behavior. It's reprinted here to show you how Hacking H(app)iness on a personal level means understanding and embracing the idea of shared value for your life.

SOCIAL RESPONSIBILITY: IT'S NOT JUST FOR BRANDS ANYMORE

It's 2015 and you're trying to get into an exclusive SoHo club. You fidget while the bouncer holds his smartphone to your face. From behind the red velvet rope, he takes a step back, his face morphing into a mask of disgust.

"You haven't done jack for anyone else in over a week?" His voice is loud enough that others in the line point their devices at you as well. "No way you're getting in here. This club is for people who give a damn about things other than themselves."

Your face burns with embarrassment as everyone calculates the accountability-based influence (ABI) score branded on your forehead like a virtual scarlet letter. Your lack of involvement means you've been ostracized by the "in-cloud."

What if your action-based reputation preceded you digitally? In one sense, brands have lived with this type of situation since the late sixties, with the advent of corporate social responsibility [CSR]. Today, social media's focus on transparency has changed the attitudes of consumers and employees regarding the modern corporation. People won't buy products from or work for organizations that aren't actively trying to change the world for good.

At what point will this lens of morality be turned on individuals, where a variation of CSR is more personal? Although, to some extent, we're already judged on our actions, we may soon be measured by the accountability metrics applied to organizations. Individuals need to understand how their actions will, quite literally, speak louder than their words.

The good news is that CSR is evolving as brands incorporate social good into their everyday business activities. Those lessons, largely based on the concept of shared value described below, provide a roadmap for individuals: how they can increase their personal value, or accountability-based influence score, while living a life that benefits the greater good.

Shared Value—Corporate Social Responsibility Permeates the Enterprise

"The idea in its purest form is that you reorient your business model around the fact that there's an intersection between where your company is trying to go and what's best for society," notes Margaret Coady, director of the Committee Encouraging Corporate Philanthropy. "[CSR] is not a department or a job title—it's a strategy for a firm to be successful over the next few decades."

This model of sustainable value creation, similar to Michael Porter's idea of shared value, moves CSR out of the realm of pure philanthropy and refocuses the notion of value based on overarching business strategy. For example, Coady cites that Western Union has created policy round tables that discuss immigration reform. At first, this seems odd for a company focused largely on wiring money. But, as Coady points out, many of Western Union's customers are successful immigrants who wire money home to their families. "[While immigration reform] may be a controversial issue for a company to engage in, it's central to the core customer of Western Union. It's a social issue that's also a business issue."

In other words, the stuff you do that helps the world will help your bottom line. That same logic applies to individuals.

As online influence metrics evolve, you'll be able to increase your ABI score as you pursue everyday career and lifestyle actions—all while making the world a better place.

Supportive Accountability—Weight Watchers Members "Lose for Good"

"The [Lose for Good] campaign actually started as an idea from a local leader," explains Cheryl Callan, chief marketing officer for Weight Watchers. "She was looking for ways to inspire members and told them to stack up piles of food equivalent to the weight they lost. It may not feel like that big of a deal when you've lost ten pounds, but when you see the food stacked up that you didn't eat, it's very motivating." When members then brought the food to local pantries, they got to see how their efforts directly help others.

Since the campaign launched in 2008, Weight Watchers members and online subscribers have lost almost twelve million pounds. The brand has donated nearly three million dollars to charitable partners that provide children and families with access to nutritious food. The campaign was "not just the CEO who had a cause and somebody wrote a check," notes Callan. By correlating the success of members losing weight to the good they could affect, Weight Watchers utilized shared value as a core business strategy.

For members, the Weight Watchers accountability model meant supporting others while encouraging more personal success.

Digital Checks and Balances—LinkedIn's "Volunteer Experience and Causes"

"What happens when someone ticks off ten causes they don't care about?" According to Nicole Williams, connection director at LinkedIn, the social community compensates using its system of checks and balances . . . "So if you say you're involved and you're not, people will call you out. Influence almost holds you accountable."

A problem for many people is the notion of touting or flaunting their volunteer experiences, who may find it cheesy and insincere. Williams urges those people to reconsider. "The more you put out there, the more that people will want to volunteer as well."

New research from LinkedIn shows that one out of every five hiring managers in the U.S. hired a candidate because of her volunteer work experience. Giving back can be the determining factor between two similar candidates.

A person can volunteer for organizations that reflect their overall passions as a way to get experience for a future career. "Volunteering used to only be about give, give, give," says Williams. "Now it's more cyclical—people are thinking about where they can contribute that's also in the best interest of their careers, so they don't get burned out."

Checks and balances aren't just about accountability. LinkedIn demonstrates the idea of shared value for an individual, a model for users to leverage their volunteerism for the greater good and for their own personal good.

Constructive Competition—Recyclebank and Rewards

"Influence needs to shape up and be better defined," says Samantha Skey, chief revenue officer for Recyclebank, an organization that rewards their 3.3 million members for taking environmental action with deals from more than three thousand businesses.

"How much does online influence impact real-world action?" asks Skey. "Is the credibility someone earns for social action online a true representation of what they do in the real world? Correlating action to intent is something we're working on measuring."

Recyclebank measures using its "Eco-IQ" score, which

identifies and helps change mainstream awareness around various environmental issues. Eco-IQ lets Recyclebank engage with a mainstream audience that may not otherwise address sustainability from an angle of environmental concern. Eco-IQ encourages peer recognition for affecting positive change in one's everyday lifestyle. "We're starting to see interest from individuals in promoting their own actions and the good they're doing, because they enjoy the pat on the back that comes from being an 'Ambassador of Green.'"

Some might fear that the reward incentive clouds a person's genuine motives. But in the case of Recyclebank, the "friends justify the means"—the message needs to reach the masses, and celebratory dynamics work better than threats. Evolving your Eco-IQ in public means shared value for everyone involved.

Positive Profiling—The Evolution of "Klout Style"

"If you work for a certain cause, it's easy for us to see the rest of your identity. So people would notice the difference between a guy who volunteered once versus someone who is really passionate . . . on an ongoing basis," says Joe Fernandez, founder of Klout.

Fernandez said there weren't immediate plans for a "Klout for Good" score per se, but the service's existing "Klout Style" feature has the potential to include a social good metric. For instance, two people have Klout scores of fifty. One influences primarily via sharing links, so Klout calls him a curator. The other user influences via talking, so she's a conversationalist. "I can definitely see expanding that to identify the type of person who is super responsible and cares about the greater good."

In this way, people perceived by the social community as bringing positive change could more overtly receive shared value of their own. Therefore, the notion of "social access" described above may not be as hypothetical as it seems.

Evolved Capitalism: IBM and the Inevitability of Shared Value

"You can't construct a carefully shaped public image which is out of synch with who you actually are," says Mike Wing, VP of IBM's strategic and executive communication [department]. Seen as the father of IBM's Smarter Planet campaign, Wing is an expert on the "Internet of Things" and posits about the future of influence, when social good will become hyper-digital and location-based.

"Data ubiquity and potential transparency will take the qualities of what people are seeing in social media several powers higher," he predicts, "and we will increasingly be making judgments based more on behavior than self-presentation."

The prediction may seem dystopian, but it's simply an evolution of our current digital existence. And as Wing points out, the bigger picture for the digital revolution will see emerging markets participating in the global arena. The two to three billion people who are now digitally connected is indicative of the fact that we've moved from a fat-tail world to a long-tail world.

In other words, the evolution of CSR via shared value is more than inevitable—it's almost passé. "CSR is a function of a previous paradigm," notes Wing. "In this context, there's no sensible way you can talk about . . . the distinction between pure wealth creation and pure philanthropy."

Accountability-Based Influence as the CSR Metric
for Individuals

While a person's actions can be perceived in a myriad of ways, various metrics from multiple sources aggregate that person's digital persona. The projects described above provide shared value for individuals and the world around them as participants increase their accountability-based influence. By encouraging the notion of ABI as a metric, people will increase their positive actions.

The moral imperative to do good has been directed toward corporations. Why not adopt their evolved idea of shared value and change the world, for good?[3]

From Takers to Givers

Give and Take: A Revolutionary Approach to Success provides an intriguing example of how the idea of shared value applies to interpersonal relationships. The book was written by Adam Grant, the youngest tenured professor at Wharton University, who describes how in professional interactions most people operate as either takers, matchers, or givers. Takers brazenly seek their own gain, while givers sacrifice their needs, many times to their own professional detriment. The majority of employees are matchers, people who feel it's important to maintain an equilibrium with colleagues in terms of social capital.

I interviewed Grant to discuss how his work reflected a shared-value mind-set. He pointed out that "productivity and profitability don't have to come at the expense of supporting other people. You can succeed in ways that lift people up as opposed to cutting them down."[4] I asked him to elaborate on this idea, having succumbed to my giver personality where others have taken advantage of my nature at work.

People think being helpful and being generous means never advocating for your own interests. Givers often will sacrifice all of their time for others, making sure they're always available. It's obvious that this type of behavior pushed to the extreme isn't sustainable. You can't succeed if you never advocate for your own interests. How to deal with this situation is to pick an interest that aligns with that of a colleague—then your time spent advocating on an issue helps advance your shared goals.[5]

I also asked Grant about the future of Hacking H(app)iness at work. I wondered how digital tools and affective sensors would apply to the giver/taker/matcher model. In response to the idea that our meetings in the future may have digital facilitators that will indicate when someone has dominated a conversation, Grant noted that "in those situations a lot of people experience something social psychologist James Pennebaker calls 'the Joy of Talking.' Most of us find that communicating our thoughts to others is a purely enjoyable learning experience and it's hard to give other people the floor."

It can be difficult to realize when we're being takers and dominating a conversation. Digital tools will be helpful reminders in the future to let others contribute to conversations where they might normally remain reticent. The unique ideas provided by people who normally remain silent will increase a company's bottom line. Shared value for interpersonal relationships will create as much benefit as it does for a company's physical supply chain.

The Balance of Well-Being

Intrinsic happiness isn't based on a momentary increase of mood. It comes from spending time with family, or practicing a craft that takes time to develop, like learning an instrument. Shared value functions in a similar fashion. While short-term gains bring temporary satisfaction, they can't sustain an organization over the long haul. Part of Hacking H(app)iness is understanding when established mind-sets aren't working so you can begin benefiting from new ideas that will bring great profit to your life.

FROM CONSUMER TO CREATOR

> Along with economists, politicians, business report-
> ers, and advocacy groups, we habitually describe
> our fellow humans as *consumers*. Of course, that
> term makes sense when applied to people wolfing
> down food and drink, but lately it has been extended
> to virtually every area of our lives . . . Until recently,
> just about everyone accepted this insidious new
> moniker, perhaps not even noticing when the term
> *consumer* began to push aside references to our-
> selves as *citizens* or simply *men and women*.[1]
>
> ANDREW BENETT

YOU'RE A CONSUMER. You, the reader. I've worked hard to establish
a relationship with you, quoting smart people and pouring my
heart out in this book. I've tried to point out that your actions, your
words, and your essence are reflected in a digital context that will
define you like never before in the future.

But fuck it. You're a consumer, I'm a consumer, we're all just
consumers. That word is a lot easier to deal with than all this technology bullshit. I'm not even a real futurist. I talk about stuff that
already exists and project a few years in the future. So I'm a pres-
entist, or a speculativist.

So let's stick with words and ideas we've become used to. I'm a
consumer and so are you. Right? I don't need to argue this point.
We're consumers.

For instance, we both know you're only invested in this book until something new comes along you want to consume. That probably happens every five seconds or so. And I'm only interested in you long enough to buy my book. Right? If I have good quotes on the jacket liner, a sexy title, and some pithy language, maybe I get lucky and you choose to consume my bit of philosophy versus buying four lattes. That's the logic of consuming, right? Comparison and choice with an onus to purchase. A *mandate*.

Yes, let's be clear: The word *consumer* comes with a mandate. You buy something. It's not a choice. Don't say the word ever again and think it's innocuous. Understand its ramifications, its deeper meanings. And realize it's being used to *define* you. The fact that you're a man, woman, old, young, live in Seattle versus Oslo, worship in a church or temple—those facts are secondary. First—*first*—you're defined as a consumer.

Consume.

Put the word in your mouth and say it slowly. It starts with a hard *C* sound, which gives it verbal power from the get-go. Then the combined *N* and *M* sounds add a lascivious undertone, an almost sexual allure that says, "You're worth this." And although it's not pronounced, the word *me* makes up the end of the word.

Say the word out loud now. Say it to your son or daughter. Look at your mom, point your finger, and say, "You're a consumer." At Starbucks with a friend, point at their coffee and say, "What did you choose to consume? Did you want to consume some more with me? Maybe next week we can come back here and do some consuming together."

Go ahead and say I'm overreacting. "It's just a term applied to people when speaking in the context of what we buy," you say. You think? Or do you think the term has shaped *why* we buy in the first place?

Of course it has. In its modern context, "consumer" is a core economic term. Someone produces something, and you consume it. This relationship implies that we're reliant on someone else to define us. Apparently we can't execute a core part of who we are until we're given the chance to *consume* something somebody else has *produced*.

Just reflect on the word for a minute. That's all I'm asking. Put the word *consumer* in your brain, take a deep breath, and let it sit there for a while. Now pretend you're looking in a mirror. How does the word *consumer* fit in that scenario? Is that the word that comes to mind when you look in your own eyes and ask, "Who am I?"

Words have power. They represent measures that have been defined by others. Using a word leads to implied acceptance of the word, which leads to forgetting how the word originated in the first place.

One of my favorite speeches in a film was delivered by Dustin Hoffman portraying comedian Lenny Bruce. It's from the movie *Lenny*, and you can watch it on YouTube.[2] The scene takes place in a smoky comedy club when John F. Kennedy was president. In the scene, Hoffman (as Bruce) accosts a number of patrons in the bar, using every racial slang term imaginable as he gets right up in people's faces. He asks the club to turn up the house lights so everyone can see one another as he continues using multiple racist terms, building the tension in the club to a boil. Eventually, when it looks as if he may actually get hit in the face, he says, "I'm trying to make a point that it's the suppression of a word that gives it its power, its violence, its viciousness." He goes on to say that if President Kennedy would go on television and use the racial slurs toward his cabinet members, the words would eventually lose their power and not mean anything anymore, and "you'd never be able to make a black kid cry because somebody called him a nigger in school."

First off, let me be clear: The racial slurs in this monologue are ignorant, hateful, and have much wider contexts than how they're used in one simple scene from a movie. My intention in this exercise is not to bandy verbiage about solely for the sake of shock value to prove a point. What I'm trying to say is words have *intent*. Shakespeare is widely known for inventing words where he felt there wasn't one in existence to express what he intended. Or here's how James 3:3 puts it: "A bit in the mouth of a horse controls the whole horse. A small rudder on a huge ship in the hands of a skilled captain sets a course in the face of the strongest winds. A word out of your mouth may seem of no account, but it can accomplish nearly anything—or destroy it!"

The English word *consumption* is from the fourteenth century and refers to tuberculosis. Consumption used to be a physical malady defined by a person's body literally wasting away. In the modern economic sense, the definition refers to people using up goods and services in order to purchase more of those goods and services. Sustainability doesn't enter into the picture. Finite resources aren't part of the equation. In the same vein that the corporate world can get caught up in quarterly profits, people get caught up in a consumptive lifestyle because that's what they've been taught is the right thing to do. The only thing to do.

But the word and what it stands for are causing a great deal of anxiety, as noted from a press release from the Association for Psychological Science about a Northwestern University study.

Money doesn't buy happiness. Neither does materialism: Research shows that people who place a high value on wealth, status, and stuff are more depressed and anxious and less sociable than those who do not. Now new research shows that materialism is not just a personal problem. It's also environmental . . . "It's become commonplace to use *consumer* as a generic term for people," [says Northwestern University psychologist Galen V. Bodenhausen] in the news or discussions of taxes, politics, or health care. If we use terms such as *Americans* or *citizens* instead, he says, "that subtle difference activates different psychological concerns."[3]

This study was conducted at Northwestern University by psychologists Galen V. Bodenhausen, Monika A. Bauer, James E. B. Wilkie, and Jung K. Kim, and featured a number of experiments, the last of which had participants deal with a hypothetical water shortage where a well was shared by four people. Identified either as "consumers" or "individuals," participants were studied to see how their adopted identities would affect their behavior toward others.

Might the collective identity as consumers—as opposed to the individual role—supersede the selfishness ordinarily stimulated by the consumer identity? No: The "consumers" rated themselves as less trusting of others to conserve water, less personally responsible, and less in partnership with the others in dealing with the crisis. The consumer status, the authors concluded, "did not unite; it divided."[4]

Lenny Bruce pointed out that words become weapons when they're made sacred by lack of use. But words also get weaponized from *overuse*—allowed in a certain context, we ignore the insidious effects the words have on our psyches.

I don't want you to think of the word *consumer* the same way ever again. I don't want to give you permission to use it to define yourself or anyone else when there are so many words to better describe your central identity.

From Consumer to Creator

In my Mashable article "The Impending Social Consequences of Augmented Reality," my friend Chris Rezendes nailed a term I'd like to propose for people in the Connected World of the future focused on holistic value versus just consumption. He said, "We're going to call people creators."[5]

It's in our nature to create. And you don't have to be creative to create. When you choose words to say in a discussion, you're creating a conversation. When you choose to talk to a cute woman at a bar, you're creating an opportunity. The act of creating means you're putting something into place that didn't exist before, something only you can bring into being.

Shifting from the word *consumer* to *creator* also has a positive effect on the people with whom you relate. As a consumer, our relationships are focused on dealing with producers. From a transactional standpoint, this makes sense. I don't produce my own vegetables, so I need to consume them from someone else. But I'd rather simply call that person a *farmer*, to denote the title and appreciation warranted by their profession.

If you're a creator versus a consumer, I think the people you deal with should be called savorers. I love the word *savor*, since by definition you can't do it quickly. In positive psychology, according to Fred Bryant and Joseph Veroff, the act of savoring means you appreciate the positive aspects of life.[6] You can't rush savoring. Instead of thinking of someone just as a consumer who ingests their way through existence, ask them what it is they create. Our work used to be about craftsmanship before the Industrial Revolution. Technology has improved how we get things done, but it doesn't mean we can't go back to former language that acknowledges people's investment in their work.

The Creator in the Connected World

Here's a scenario showing how a creator's life could look in the near future:

You're at a Starbucks waiting in line and get an IM (instant message) in your HUD (heads-up display—like Google Glass, but it covers both your eyes). A camera icon with a question mark appears in the upper-right-hand portion of your vision, meaning someone nearby wants to take a picture and you're in the shot. You IM back an automated message:

> *Hi. Looks like you want to take a picture that would feature my image. Since we're in a public space, I can't keep you from snapping. But if you use facial recognition to tag me, note that I own my own data in any format and will appear as an avatar in your picture unless you receive my written consent to use my image.*

You immediately get another IM with a money icon and a URL link to a blog. You blink to open the URL and see a series of riveting black-and-white photographs featuring people waiting in lines. The title of the blog is *This Is Your Queue*, and you think it's really cool. So you blink on the money icon and a message appears from the person requesting the picture that says:

> *I use PaySwarm to provide micropayments for anyone I feature in my images. I don't tag people's faces, and I have a computer program that constantly scans the Web to make sure other people aren't using my pictures without permission. So if you let me use your image, your visual data will be safe, and after a while maybe you'll make enough money to buy a coffee like the one you're waiting in line for right now.*

You blink twice toward the IM to accept these conditions, looking to the right using eye-tracking technology to save the URL for the blog so you can check it out in the future. You turn and smile at the photographer who has identified himself sitting nearby, appreciating the fact that your picture has become a work of art.

From Creator to Provider

The scenario I've described here isn't new in terms of picture-taking. There are millions of pictures on Facebook right now featuring people who didn't know they were being photographed. Many of them have been identified by facial recognition, and somebody is making money in some way off their images without their knowledge. These are *consumers*, dehumanized faces available for sale.

Change these people to creators, however, and their likenesses become contributions to the virtual landscape. They can't be sold without their permission. And permission is granted based on mutual appreciation.

In the scenario described above, the "consumer" being photographed has also gone one step beyond being a creator. They've become a *provider*. Their likeness has provided an opportunity for someone else to create their work. Economically speaking, the photographer could be considered a producer, and the person in line a consumer. But this exchange involved mutual respect and appreciation. While the photographer has promised to pay in the future, the person waiting in line has no guarantee this will happen. The exchange at its core wasn't transactional in nature as much as relational and based on trust.

The person in line also savored, however briefly, the photographer's blog. This act of reflecting on someone else's work and acknowledging their unique contribution to the world is profound. Whatever the technology interface, whatever the setting, this act of recognition says, "I recognize your efforts. And I see you."

Remember how I started this chapter calling you a consumer? Remember how it distanced us, putting us at arm's length? I did that to prove to you that words matter. You have more value to give to the world than as a vehicle for consumption. You're a creator, and a savorer. You can richly contribute to other people's lives while also deeply appreciating other people's worth. You can participate in shared value on a personal level and create intrinsic happiness in our Connected World.

I see you.

Be Proactive

PROMOTING PERSONAL AND PUBLIC WELL-BEING

THE CURRENCY OF
WELL-BEING IS ATTENTION.

—*Avner Offer*

THE ECONOMY OF REGARD

What are the advantages which we propose to gain by that great purpose of human life which we call bettering our condition? To be observed, to be attended to, to be taken notice of with sympathy ... and approbation, are all advantages that we can propose to derive from it.[1]

ADAM SMITH

WE ALL HAVE defining moments in our lives. One of mine involves bullies.

I lived next to my school growing up and was always riding my bike or going to the playground. I was on the monkey bars one afternoon when the group of kids who made a career of taunting me showed up. I jumped off the monkey bars and spun around, waiting for my latest dose of vitriol.

It didn't come. Tom, chief bully of the group, nodded toward the monkey bars and said, "Nice job, Havens. Can you do that again?"

I was confused—was this a form of kindness? Or maybe just recognition? Earlier in the day, Tom and the boys had surrounded me for a sadistic version of the silent treatment where they'd forcibly ignored me yet hadn't let me leave. Now they were being nice to me?

Of course I didn't buy it. But then Tom's toadies joined in on the compliments and seemed genuinely impressed. A wave of joy swept over me. Was my time of being bullied coming to an end? I didn't genuinely believe that it was, although I wanted nothing more at the time. Call it gullible or wishful thinking; I was just happy to get noticed.

One of the toadies spoke up: "Why don't you do that again?"

I scrambled up one side of the monkey bars, thrilled at the prospect of an audience. I hiked up my shorts and jumped. As I grunted my way toward the other side, I saw Tom lunge forward in my peripheral vision. Confused, I held on to the bars as he gripped my shorts and pulled them down around my ankles. The force of the pull was strong enough to pull my underwear down, exposing my naked butt.

In the middle of a hot spring day, a full moon.

I jumped off to the sound of bullies howling with laughter. Pulling up my pants with an awkward heave, I swore with rage. While I hadn't directed it at Tom, he picked up on it immediately—he was a top-notch bully in terms of technique.

"What'd you call me?"

"I didn't call you anything." Much like the classic scene from the movie *A Christmas Story*, where one kid dares another kid to stick his tongue to a frozen pole, rituals around fighting were fairly defined in my elementary school.

"You fucking swore at me, Havens. And now I'm gonna kick your ass." Tom was going straight from the playbook, so I responded with a classic counterpoint response.

"But there's five of you," I pointed out. For emphasis, I added, "And only one of me."

"Just you and me, fat boy," Tom replied, the other bullies moving aside like a bully ballet. "Just you and me."

There it was. The trap had been set, and I'd been forced to spring it. I had one last card left to play, however. When you get bullied a lot, being called a chicken doesn't really mean anything, since you get mocked all the time. So I decided to feign gallantry and walk away.

"I'm not gonna fight you, man," I said, and turned to leave.

I took about three steps before I heard someone run up behind me. I began to turn around, but Tom was quicker than I was, and he jumped on my back. His friends howled with laughter as Tom repeatedly punched my face.

I'm not sure how I got Tom off my back and turned around. But I did, and I put him in a piledriver hold—I gripped him around his back like a wrestler while his head was facing the ground. It's a pretty disorienting position to be in, and as I started running Tom backward, he shouted for me to stop. So I did. But as I removed my arms from his back and he raised his head, I brought my knee up into his face hard, three times. On the last impact, I felt something snap and my leg became covered in blood.

I had broken Tom's nose. Didn't mean to. Not proud of it. But I won't lie: It felt *really good*.

A central tenet of Hacking H(app)iness is acknowledging that you can't fully appreciate what makes you happy until you've lived through experiences that make you miserable. Extremes provide measures of comparison.

After I broke Tom's nose, I went home and got him paper towels. When I came back to the playground, he was sitting with his head tipped back, fellow bullies spread around him in a semicircle. I gave him the towels and asked if he was okay. He said he was fine.

And then the strangest thing happened.

I stood there and we just chatted for a few minutes. We talked about our fight like we had watched it on TV. Tom actually complimented me on my piledriver move. The other bullies chimed in as well. And for a fleeting, blissful moment—

I was one of the guys.

Apparently Tom knew he had broken two unspoken playground rules:

1. It's chickenshit to jump a guy from behind.
2. Sucker punch a rage-infused kid you've bullied for months at
   your own risk.

Tom's ignoring our rules meant the veneer of our playground
world had been broken. And I had never appreciated being paid
attention to more than I did in that moment.

Sadly, the next day, we resumed our usual roles, and I sank into
my lonely rituals once again. Fortunately, the abject misery I
experienced at school was offset by an amazingly wise and loving
mother who gave me advice that guided me through elementary
school and beyond: "There's always someone worse off than you.
Find them, help them, and you'll feel better."

Her solution worked. The desire to lessen one's suffering is a
great motivator for empathy. Ironically, this push to help others
was fueled by a powerful need to help myself. It wasn't an act of
selflessness as much as desperation. It was heartfelt, but, by definition, I needed others to remove my sense of isolation.

What I discovered through my playground experiences was
what economist Avner Offer calls "the economy of regard."[2] I learned
firsthand about supply and demand regarding attention. Being
ignored proactively deteriorated my well-being. It's why I had so
much anger built up and why Tom's septum is probably still devi-
ated to this day.

Adam Smith, father of modern economics, first posited the idea
of the efficiency of an impersonal market in his seminal treatise of
modern economics, *The Wealth of Nations*. Typically referred to as
"the invisible hand," Smith's central idea in the book claimed that
people's efforts to maximize personal gain in a free market benefit
society.

Smith's first work, *The Theory of Moral Sentiments*, seems to contradict this idea of an impersonal market, however. Here's how Wikipedia summarizes this point:

> In the work, Smith critically examines the moral thinking of his time, and suggests that conscience arises from social relationships. His goal in writing the work was to explain the source of mankind's ability to form moral judgments, in spite of man's natural inclinations toward self-interest. Smith proposes a theory of sympathy, in which the act of observing others makes people aware of themselves and the morality of their own behavior.[3]

I never thought of my wanting to help other people as economic in nature. But observing other people's suffering did make me try to help them, which provided a form of positive social exchange.

In his book *The Challenge of Affluence: Self-Control and Well-Being in the United States and Britain since 1950*, Avner Offer notes that, in *The Theory of Moral Sentiments*, Adam Smith "described the purpose of economic activity as the acquisition of regard."[4] Regard is also known as approbation, an appreciation between two people that can come in different forms, such as attention, respect, kinship, and acceptance. I interviewed Offer and asked him to elaborate on these ideas.

> We try to maximize our well-being by being worthy of other people's approbation, and we earn it by giving it away. It's a social exchange in a relationship of equals. A basic resource we all have is self-worth. But you can't feel self-worth just through bootstrapping—we need the validation of others. That's the primitive core of the relationship, the idea of reciprocity.[5]

Reciprocity becomes an essential practice in society because it's impossible for people to function just through impersonal transactions. *Social* economics means you can't always use money for exchange. Traditions like hospitality dictate the giving of gifts based on building a relationship. You bring a bottle of wine to a party or you feel like a jerk. And if you forget to bring something, you don't offer the hostess twenty bucks. You try to make up for your loss of social capital by repaying her with a kindness at some point in the future.

This economy of regard equates to shared value in the sense that people only feel contentment if everyone benefits. People need to be observant of one another. They need to see one another. Regard is a mutual process where value is only created when two people actively experience each other in real time.

This face-to-face nature of regard has begun to erode in the wake of digital technology. Too much time spent staring at screens can have physiological repercussions on our bodies, where the plasticity in our faces that helps us smile can begin to atrophy. Barbara L. Fredrickson, the Kenan Distinguished Professor of Psychology and principal investigator of the Positive Emotions and Psychophysiology Lab at the University of North Carolina, Chapel Hill, and author of *Love 2.0: How Our Supreme Emotion Affects Everything We Feel, Think, Do, and Become*, explains this phenomenon in her *New York Times* article "Your Phone vs. Your Heart." Her research shows that when people take each other in, part of how they share a smile or a laugh is physiological in nature. Mirror neurons create micromoments that cannot be exchanged if both parties are not fully present in a moment. These precious exchanges build capacity for empathy and even improve our overall health.[6]

Apparently the economy of regard doesn't apply to screens. Oxford scholar Robin Dunbar, an English anthropologist who is an expert on real-world social networks, has also famously noted that, no matter who you are or where you live, no person can hold more than 150 people in their active social circle.[7] Apparently attention is a scarce resource—it can only be paid to a finite number of people.

The Pursuit of Provision

I have good news. There's always going to be someone who needs your help. And in an economy of regard, if you pay attention more than you receive it, you'll be providing a valued benefit to the people in your life. In the Connected World of augmented reality, it's going to be easier than ever to only see the people we want to see. We can get caught up in amassing followers and feeding our egos. We can spend all of our attention on ourselves.

Or we can spread the wealth and look outward. Shifting from a consumer to a provider mind-set, we can give Big Data direction and analyze which needs around us can be met with the time and skills we have available. Civic engagement doesn't have to be just a duty. Paying attention to others feeds our inherent need for approbation in an economy of regard.

In an article for the *Atlantic*, Kathleen Kennedy Townsend, daughter of Robert Kennedy, recounts a quote from her father, who was asked by British media personality David Frost about the purpose of life. Kennedy's answer was remarkably simple, saying if you have enough to eat and survive, your focus should be to help others who don't have those advantages, and "you can always find someone that has a more difficult time than you do."[8]

Townsend's article also points out how her father critiqued the idea of the gross domestic product (GDP) as a measure of national well-being, and how newer metrics like Bhutan's Gross National Happiness are expanding how the world views value creation. Referring to the quote from the Declaration of Independence, "life, liberty, and the pursuit of happiness," Townsend points out that the Founding Fathers were, by and large, wealthy and could have avoided helping their communities beyond their core political obligations. But they "believed that you attained happiness, not merely through the goods you accumulated . . . but through the good that you did in public."[9]

I interviewed Townsend, who served as lieutenant governor of Maryland from 1995 to 2003, where as part of her work she was the founder and executive director of the Maryland Student Service Alliance that made Maryland the first state in the country to include a high school community service requirement—an act mirroring the Founding Fathers' focus on doing good for others. I asked her how she felt the GDP and a focus on individualism had affected modern Americans' views on happiness.

> People have been told to think about themselves. And in terms of what do we mean by "happiness" and how do we become happy—people think that simply getting more for themselves makes them happy. I think that's the message of our consumer society.[10]

Community comes at a price. By definition, it involves coming together with other people around shared values or needs. Pursuing happiness is a public function as well as a private one. It's your choice to stay in isolation and only utilize your skills to build up individual wealth. But why keep other people from experiencing your awesomeness? Spread the love.

According to a great number of positive psychologists, you'll increase your happiness by utilizing skills that you feel bring you the greatest meaning or by participating in an altruistically focused action. The phrase "pursuit of happiness" indicates an ongoing journey, a series of events versus a finite state. Long-term happiness comes as a result of actions taken versus a reliance on momentary shifts of mood.

Ask yourself if you're contributing to the economy of regard or spending all your time reflecting just on yourself. If, as Avner Offer says, "the currency of well-being is attention," why not pay it forward and try to help someone else get happy and see what that does for you?

POSITIVE PSYCHOLOGY

Most of the shadows of this life are caused by standing in one's own sunshine.

RALPH WALDO EMERSON

MY DAD WAS a psychiatrist for more than forty years. This was the first question he'd ask all of his patients:

DR. HAVENS: Do you watch the eleven o'clock evening news?
PATIENT: Yes, I do.
DR. HAVENS: Stop watching the eleven o'clock evening news.

This may sound simplistic, but it echoes a core idea of a field of science known as positive psychology—to get happier, stop focusing on what makes you miserable. It's no secret that, to get ratings, news shows feature some of the most horrific events that have happened throughout the day. You don't have to avoid hearing about them, but you can plan on optimum times to focus on these types of messages during your day when they won't affect you as negatively as other times. (Newsflash: Right before you go to bed is not one of those optimum times.)

Growing up with a psychiatrist for a father made a lot of kid experiences different for me than most. For instance, Take Your Child to Work Day wasn't really an option for the son of a psychiatrist in the early seventies. Since my dad worked in a private practice, the only people who called him were his answering service and patients who got our home number. One time a patient called when I was about ten, and I answered the phone:

ME: Hello, Havens residence.
PATIENT: Is this John or Andy [my brother]?
ME: John.
PATIENT: John, I'm coming over to your house to kill your father.

Well, that sucked. After dropping the phone and running to tell my dad about the call, he calmly responded, "Was it a man? With a gravelly kind of voice?" I told him yes. "Don't worry about it," he said. "That's Richard. He's harmless. He's just lonely."

Looking at my dad, reassuring me with a smile, I realized how heroic his work was. It's hard to experience other people's pain at point-blank range. The raw emotions most of us deal with on a sporadic basis were my dad's daily bread. The fact that he got paid to do it was inconsequential; he would have done it for free. In fact, many times he would work with patients even if they couldn't pay.

Note I said he would "work" with patients. This language is key. If you go see a therapist, you're making a bold and admirable step. You're seeking help. But once you get there, be ready to *work*. As a writer and actor, I've had the benefit of years of training around introspection. It doesn't mean I can control my emotions, mind you; it just means I understand the need to measure how I feel and behave where I feel I need to improve.

I'm saying this because I don't want to bullshit you in terms of Hacking H(app)iness. Taking measure of your life is hard. It should be. Otherwise it wouldn't be rewarding. You're worth the effort of deep self-reflection. That way, you get to discover what makes you tick. You'll also learn what actions you can take to amplify the positive things in your life while decreasing the negative.

Here's some more straight shooting for you: The majority of science around positive psychology shows that the mood of elevated feelings often associated with ephemeral happiness is fleeting. This type of happiness is called hedonic happiness (same root word as *hedonism*) and stems from short-lived experiences we get used to quickly. This doesn't diminish the joy you'll feel at these moments, like when you get a raise or buy a new car. But within a week or two, you may find yourself needing another rush from a similar type of experience, and you may get caught up on what's called a "hedonic treadmill." This means you'll adjust, or habituate, your core level of happiness to this new event. Pretty soon it won't bring the same pleasure it did at first.

This is the type of emotional experience most of us associate with happiness—the rush of romantic love, the thrill of an exotic vacation. These are normal, valid, and frankly awesome emotions to have. But you can prolong another type of happiness known as eudaimonic happiness by focusing on things that bring you intrinsic rewards. *Eudaimonia* is a word coined by Aristotle and is often translated as "human flourishing." Flourishing implies a long-term state of being versus momentary mood. This is why, in academic or scientific discussions, people often use "well-being" instead of "happiness" to discuss these issues. "Happiness" in these contexts can be construed as mood, or the narcissistic pursuit of pleasure for pleasure's sake alone. Eudaimonia, by contrast, refers to the highest human good one can achieve. By definition, seeking eudaimonic well-being implies an outward focus in order to flourish within. For the Greeks, this also meant interaction within one's polis, or city, as helping others in the community was a way to increase long-term and intrinsic well-being.

Here's some good news: I'm not going to give you a bullshitian how-to set of rules to follow in this book that will "guarantee your happiness." The sobering, yet also good, news: I'm going to walk you through an amazing set of scientifically proven theories showing how you can work to identify the things that bring you meaning, and how you can amplify them to increase long-term well-being. But first you have to be willing to do the work.

It's like the old psychiatrist joke my dad loved to tell:

DAD: How many psychiatrists does it take to change a lightbulb?
ME: How many?
DAD: Two. One to get the ladder, and the other to ask the lightbulb if he really wants to change.

The Science and the Sacrifice

Do you want to change? Do you want to work to improve your level of happiness? It's okay if initially you think that this idea is a load of hooey. But don't let that stop you from experimenting with the ideas you'll discover from some of the leading scientists around the world. Rather than a self-help formula, however, a lot of what positive psychology reveals feels like common sense. It's just common sense backed by science.

A big part of Hacking H(app)iness is about taking action to improve your well-being, and I want to provide some proactive tools to get you started right away.

Here's my first recommendation: Go to Happify (www.happify .com) and sign up to try their site. Their five-part STAGE framework (Savor, Thank, Aspire, Give, Empathize) is based on the science of positive psychology. They point out on their site that "recent scientific breakthroughs reveal that happiness is a skill within your

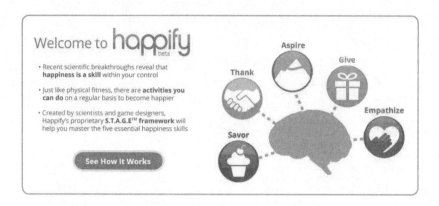

control." This is backed up by a number of other psychologists and scientists. Like exercising your body, you can exercise areas of your life that will increase happiness. No time like the present—sacrifice a little bit of time and see how happy you are with your results.

When you go to the site to sign up, you'll be asked to fill out a short survey (took me three minutes) to assess your happiness level, a number to show where you're optimized and where you could improve. Then you'll be given a number of tracks to choose from to help begin the work of improving your happiness. I chose the "strengthen your friendships" track, because even though I'm a very outgoing person, I'm also a homebody as a writer. I was given the "thanks for being awesome" task that asked me to write down three things I appreciated about my best friend. Here's what I wrote:

- She listens to the details of my work even when I know she hasn't always told me about her day.
- She is always thinking ahead for what's best for our kids and our family.
- She loves me for who I am. That is not always easy for *me* to do.

I felt pleasure writing those words. It took about sixty seconds to think about how awesome my wife is, and I felt a renewed sense of blessing that she's in my life.

I paused. I reflected. I remembered how freaking lucky I am. I got happier.

One of my favorite aspects of Happify is how, after you finish a task, you can click on their "Why It Works" button to read about the science behind the activity you've just done. Here's what was listed after my "thanks for being awesome" exercise:

> In a study conducted by Drs. Martin Seligman, Tracy Steen, and Christopher Peterson, a group of people was asked to practice this gratitude exercise every day for one week. Even though the exercise lasted just one week, at the one-month follow-up, participants were happier and less depressed than they had been at baseline, and they stayed happier and less depressed at the three- and six-month follow-ups. This practice primes our mind for gratitude and helps overcome the brain's natural "negativity bias," a phenomenon by which we are wired to give more weight to negative rather than positive experiences or other kinds of information.[1]

The negativity bias, by the way, is why my dad told people not to watch the news—you have to train yourself against your brain's proclivity to heed negativity. Apparently it's a remnant from caveman days when being aware of negative things like "that mammoth looks angry" or "neighbor Grog has ax aimed at my head" helped keep us alive. These days, that ancient bias means we feed off distressing news or even gossip.

Let it go, people. Mammoths are extinct. It's time to focus on the positive.

A Complement to What's Come Before

The science of positive psychology has been Hacking H(app)iness for over a decade now, helping people focus on ways to improve their well-being versus simply removing pain. Note the field is not trying to replace traditional psychoanalysis, but to complement it. Utilizing the scientific method in analyzing human behavior, the field seeks to prove that focusing only on people's disorders could lead to an incomplete view of their condition. Here's how the International Positive Psychology Association answers the question "Is positive psychology an abandoning or rejection of the rest of psychology?":

> In a word, no. [The] consequence of this focus on psychological problems, however, is that psychology has little to say about what makes life most worth living. Positive psychology proposes to correct this imbalance by focusing on strengths as well as weaknesses, on building the best things in life as well as repairing the worst. It asserts that human goodness and excellence is just as authentic as distress and disorder, that life entails more than the undoing of problems.[2]

In a similar fashion, Gross National Happiness and other metrics of well-being around the world are widening people's perspectives around measuring value. While measuring wealth is an important metric, it isn't the only determinant of happiness or well-being. In his book *Flourish: A Visionary New Understanding of Happiness and Well-being*, Martin Seligman agrees with this sentiment, noting that if all that is being measured by GDP is money, policy will be focused only on getting more money. By measuring well-being, policy will reflect a wider scope of measures beyond fiscal wealth.

In terms of your life, if all you measure is the negative, guess what you'll focus on?

In terms of your digital life, if your main priority is increasing your online influence, you'll discover the hedonic treadmill firsthand (literally—your thumbs are probably tired from posting on Facebook). If you measure your life only by your Klout score, you'll never achieve long-lasting happiness.

To increase our well-being, we need to look beyond ourselves.

Martin Seligman and PERMA

A great way to introduce yourself to the work of Martin Seligman is to listen to his talk about the state of psychology from a TED Conference in 2004.[3] In about twenty minutes, Seligman walks through specifics on the nature of traditional psychoanalysis and how positive psychology as a science is now complementing the study and improvement of well-being. Seligman is called the father of positive psychology, although there are others, like Mihaly Csikszentmihalyi and Sonja Lyubomirsky, who are also credited with creating the field (we'll discuss them in future chapters on the concepts of flow and altruism).

Here's how Wikipedia defines positive psychology:

> Positive psychology is a recent branch of psychology whose purpose was summed up in 1998 by Martin Seligman and Mihaly Csikszentmihalyi: "We believe that a psychology of positive human functioning will arise, which achieves a scientific understanding and effective interventions to build thriving individuals, families, and communities." Positive psychologists seek "to find and nurture genius and talent" and "to make normal life more fulfilling," rather than merely treating mental illness.[4]

In Seligman's recent book *Flourish,* he discusses his idea of PERMA,[5] or the five measurable elements of well-being (versus happiness), the primary focus for positive psychology.

POSITIVE EMOTION—You can act on this by being grateful, either by journaling or telling someone else how they improve your life.

ENGAGEMENT—You can act on this by identifying the core skills you think you were built to accomplish. This is the idea of discovering your "flow," a term coined by Mihaly Csikszentmihalyi that we'll discuss in a future chapter. When you achieve a state of flow, you don't feel anything in the moment, since you're so deeply involved in what you're doing. It's after a task is completed and you reflect on it that you feel a deep sense of satisfaction and accomplishment.

RELATIONSHIPS—You can act on relationships, but since they include other people, they become multifaceted but necessary aspects of improving well-being.

MEANING—You can act on this by serving a purpose bigger than your own fulfillment. Like the Greek focus in eudaimonia about civic engagement, meaning is created from outside of yourself, not solely from within.

ACHIEVEMENT—You can act on this when you pursue success, accomplishment, and mastery for their own sakes, even if they bring no positive emotion or increase in positive relationships. Similar to the idea of flow, people often pursue achievement in things like sports for the sheer joy of participating in that activity. As an example of this, in *Flourish*, Seligman quotes the actor playing famous Olympic runner Eric Liddell in the film *Chariots of Fire*: "I believe that God made me for a purpose . . . But he also made me fast, and when I run, I feel His pleasure."[6]

In *Flourish*, Seligman points out that PERMA and well-being are constructs, where happiness is a "real thing." In positive psychology as well as economics, measuring happiness is often done by asking people to fill out a survey question focused on "life satisfaction." Typically this contains either a seven- or ten-point scale, where one indicates low life satisfaction and seven or ten indicates high life satisfaction. These scales are a useful tool because they ask people to provide their subjective perspective on how they feel about a certain experience or aspect of their lives. While they may be affected by survey bias, a term meaning they know they're being asked a question and may change their answer based on that awareness, they're providing a truth that nobody can deny. Subjective in nature, employing a large-scale survey about citizens' life satisfaction around a certain issue lets a government discover the answer to "How are we doing in this area?"

So while a happiness-focused survey can measure a singular subjective question with rigor, well-being, as Seligman defines it, contains multiple measurable elements versus one overarching answer. Each element contributes to well-being, but doesn't define it as a whole.

Seligman is the director of the Positive Psychology Center at the University of Pennsylvania, and his site, Authentic Happiness, features a number of helpful surveys you can fill out to learn more about your attributes as they pertain to PERMA and other positive psychology–related elements. I found the Grit Survey particularly helpful. It has twenty-two questions and took about five minutes to fill out. According to the site, "Grit is perseverance and passion for long-term goals. Our research suggests that grittier individuals accomplish very difficult challenges." I got a three out of five, which I found to be quite interesting. While I thought my score might have been higher, I tried to answer the questions as truthfully as possible, a process I found enlightening.

A final word from Seligman about the nature of positive relationships in relation to PERMA: "Very little that is positive is solitary," he notes in *Flourish*. He quotes his friend Stephen Post, professor of medical humanities at State University of New York at Stony Brook, relating a story about his mother. When Stephen, as a boy, looked flustered or upset, his mother would encourage him to "go out and help someone." As it turns out, this piece of maternal wisdom has empirical backing, as Seligman notes, "We scientists have found that doing a kindness produces the single more reliable momentary increase in well-being of any exercise we have tested."[7]

As my mom said this same thing to me, apparently I was raised by two experts in psychology, not just one. I'm a more positive person because of it.

Positive Psychology at Work

"With the science of happiness and positive psychology, we're focusing on what's right with us and fine-tuning those things—like our sense of progress, control, connectedness, and purpose—to become happier people," noted Jenn Lim.[8] Jenn is the CEO and chief happiness officer at Delivering Happiness at Work, the workplace consultancy created by Tony Hsieh, CEO of Zappos.com and author of the best seller *Delivering Happiness: A Path to Profits, Passion, and Purpose*. The company is one of a growing group of organizations utilizing principles of positive psychology to inspire increased business value along with inspired employees.

In 2002, Gallup quantified a link between employee feelings and ROI, reporting that lost productivity due to employee disengagement costs more than $350 billion in the United States every year.[9] In recent years, there's been a shift to quantify positive emotions or happiness in the workplace as a way to increase revenue, rather than simply worry about loss due to employee disengagement.

An interesting aspect of the study of happiness at work has to do with global cultural attitudes of emotions and their place in the enterprise. I interviewed Marise Schot, a concept developer and head of the Happiness Lab at the Waag Society in Amsterdam who also founded her own design studio, to gauge how she felt U.S. attitudes toward emotion at work were different from those of the Dutch.

My experience is that in the U.S.A. the existence of emotions is less present in daily life, and also less accepted than in the Netherlands. What I noticed was that food was used to treat yourself or as a way to take a break (the stroll to the Starbucks, lunch meeting with colleagues) or when you think you deserved it. For us Europeans, this idea of having food manage your moods and needs was not something that we recognized. But it makes perfect sense as you relate this with the American dream, where you are expected to work hard in order to become successful—there is no room for emotions.[10]

I find this fascinating, that the perception of American culture could be that people in the United States don't have time for emotions regarding work, or that an excuse might be required to go off-site to express one's feelings. Where food in Marise's example is the instigator of expression, sensors or other technology aligned with consulting practices like the ones offered by Delivering Happiness appear to be providing permission for Americans and other workers to identify and benefit from positive emotions embraced within the workplace. And while the idea of quantifying happiness in the enterprise may seem fluffy at first, Jenn Lim points out this phase of doubt will pass:

> There's always going to be naysayers, but now that we've developed ways to tie workplace benefits back to scientific, measured happiness, even they can't deny the correlation between happier employees, happier customers, and more successful long-term sustainable business. In five to ten years, happiness in the workplace won't be a novel idea— it'll be an economically proven and understood model that organizations will use as a way to ensure long-term sustainable and profitable brands. In an even shorter time, more individuals will recognize that happiness should be prioritized both at home and at work.[11]

Kristine Maudal is a partner and CFO (chief fun officer) at Brainwells, an innovation consultancy based in Oslo, Norway, helping companies foster happier and more productive workplaces and focusing on a "return on involvement" versus just standard ROI. I interviewed Kristine on the subject of happiness at work, and she noted the importance of being able to measure progress based on employee engagement.

> I do definitely think that happiness at work can be measured and improved. But it is important to define what we mean by happiness. Scholars, researchers, consultants, press, everyone is talking about the importance of work-life happiness and satisfaction, but only a few know how to create it. What we know is that people really like to be seen, heard, and involved. That makes them happy. And engaged. Engaged people make better work. That creates happy leaders. It is a good circle.[12]

A final point on the idea of positive psychology at work: Whatever cultural biases we may have, the fact remains that adults spend the majority of their lives in an office or work setting. When fifty to sixty hours a week (minimum) are spent at work, it's time we recognized that not focusing on creating well-being and happiness at our jobs means we're ignoring a large part of our emotional life for the decades we're in the workforce. The organizations that embrace methodologies to leverage positive well-being and happiness for their employees are certainly more likely to see benefits in the future than the ones that don't.

Compassion Is Catching

There's a debate among scientists about human nature regarding selfishness. Are we wired only to think about ourselves? It makes sense to think that, in an evolutionary process, helping others probably wouldn't be the best way to keep your own species alive. But a good deal of science in the field of positive psychology has revealed how compassion may be hardwired into us via the neurons and hormones that are a part of our brains.

Greater Good is a website and publication created by the Greater Good Science Center (GGSC) based at the University of California, Berkeley. The GGSC's mission is to "study the psychology, sociology, and neuroscience of well-being, and [teach] skills that foster a thriving, resilient, and compassionate society." In his article "The Compassionate Instinct" for the site, Dacher Keltner provides a number of scientific studies documenting altruism and compassion, including research conducted at Emory University:

> In other research by Emory University neuroscientists James Rilling and Gregory Berns, participants were given the chance to help someone else while their brain activity was recorded. Helping others triggered activity in the caudate nucleus and anterior cingulate, portions of the brain that turn on when people receive rewards or experience pleasure. This is a rather remarkable finding: helping others brings the same pleasure we get from the gratification of personal desire.[13]

The article also described the presence of a hormone known as oxytocin in our bodies that floats through our bloodstream. Keltner conducted a number of studies and found that when people perform behaviors associated with compassion (warm smiles, friendly hand gestures), their bodies produced more oxytocin. The suggestion of this behavior, as Keltner points out, is that "compassion may be self-perpetuating: Being compassionate causes a chemical reaction in the body that motivates us to be even more compassionate."[14]

I first learned about oxytocin and its relation to compassion

when I interviewed[15] filmmaker, publisher, and workshop producer Eiji Han Shimizu. Shimizu is on the advisory committee for the
H(app)athon Project that I founded and has a unique program that
combines Zen meditation with entertainment. He was also a producer for the *Happy* movie.

One of the best ways to learn a majority of the newest ideas
around positive psychology is to watch this film. For a number of
months in 2013, it was the highest-rated documentary on iTunes,
and has won more than a dozen awards to date. Featuring multiple
interviews from leading psychologists and other experts, its power
lies primarily in the interviews of people from around the world
and their attitudes toward happiness in their own lives.

Shimizu had a thriving career in Tokyo before working on
Happy. But in our interview, he related that success in business
wasn't helping him improve his well-being. In fact, the more successful he became, the more stress he felt. Sadly, as the documentary points out, Japan has, for many years, had the highest suicide
rate of any developed country. Stress is at an all-time high as many
young men and women seek to increase their productivity and
wealth above all else. The cost for this singular focus has been
alarmingly high.

Leaving Japan to pursue work on the film changed Shimizu's
life. *Happy* took a number of years to create, and now that it's been
released, Shimizu is leading workshops to help others discover
and foster their own well-being. In our interview, I asked Shimizu
what he thought was the most surprising thing he had learned
while working on the film.

> The most surprising thing to me was that we'd been com-
> missioned to make a film about happiness, but what we
> ended up making was a documentary on compassion. Af-
> ter interviewing a number of scientists, they verified that
> having a compassionate mind-set is the best booster of

happiness. Again and again, science has verified the strong correlation between happiness and the good heart.

This correlation is based on the discovery of mirror neurons and how they relate to oxytocin. Essentially, oxytocin is released when you are kind to someone else, or even when you see someone do a kindness for someone else. The basic idea is that you can feel a sense of compassion in the process even if you're just witnessing it.

That's why I think we human beings have survived for so long, along with our intellect. It's not just about survival of the fittest. Survival involves the intellect, but compassion plays an equal role in the process.[16]

How encouraging to know that even witnessing acts of compassion can increase physical changes in our minds and bodies that increase our well-being. These can be experienced to a certain degree in digital realms, although face-to-face sightings[17] provide more lasting results. Seeking to flourish by looking for the positive, instead of subjecting ourselves to the negative, has scientific basis in positive psychology. Looking within to examine what brings us meaning and outward to learn from or help others is a path that can lead to increased happiness.

But we do have to look.

15

FLOW

Don't aim at success—the more you aim at it and make it a target, the more you are going to miss it. For success, like happiness, cannot be pursued; it must ensue . . . as the unintended side effect of one's personal dedication to a course greater than oneself.[1]

VIKTOR FRANKL

I'M PRETTY SURE this is how my tombstone is going to read:

John C. Havens
1969–(TBD)
Loving Husband and Father
Kick-ass harmonica player

I'd be okay with this. I've played harmonica since high school, and many of the best experiences in my life have revolved around music.[2] When I'm onstage, I enter a zone of blissful ignorance where all I'm experiencing is the music in the moment. A lot of harmonica playing involves deep breathing techniques, so playing a two-hour gig is essentially an elongated meditative exercise.

Flow: The Psychology of Optimal Experience by Mihaly Csik-szentmihalyi is one of the seminal books in the positive psychology lexicon and has been a national best seller since it was first published in 1990. Csikszentmihalyi is the founder and codirector of the Quality of Life Research Center at Claremont Graduate University and, after twenty-five years of research in the field of psychology, made a realization that would guide the rest of his life's work:

> What I "discovered" was that happiness is not something that happens. It is not the result of good fortune or random chance. It is not something that money can buy or power command. It does not depend on outside events, but, rather, on how we interpret them. Happiness, in fact, is a condition that must be prepared for, cultivated, and defended privately by each person.[3]

Long-term happiness is cultivated. It doesn't simply arrive. Like the Founding Fathers noted, it's the pursuit of happiness that will bring us greatest meaning when we strive to accomplish a goal tailored to who we are. In my case, music is where I achieve a state of flow, or what Csikszentmihalyi also refers to as "optimal experience." Flow isn't always pleasant—athletes may be in a physical state of agony while achieving optimal experience. But it's this type of almost insurmountable challenge that brings deep satisfaction: "The best moments usually occur when a person's body or mind is stretched to its limits in a voluntary effort to accomplish something difficult and worthwhile."[4]

In a 1997 article for *Psychology Today*, "Finding Flow," Csikszentmihalyi reviewed a number of central premises of his book while also giving examples for how to find flow in different parts of our lives. He also describes a technique he created for sampling survey participants called the experience sampling method, or

ESM, that laid the groundwork for many apps within the quantified self movement. Developed at the University of Chicago in the early 1970s, ESM provides a "virtual filmstrip of a person's daily activities and experiences."[5] For ESM, Csikszentmihalyi and his team utilized electronic pagers that would collect information from survey participants throughout the day, including who they were with, what they were doing, and their state of consciousness. The team collected over seventy thousand pages from more than two thousand participants.

A modern example of this pager method can be found in the Memoto camera, a small wearable camera that takes pictures multiple times throughout the day. A form of quantified self in terms of tracking behavior, the camera was also created for lifelogging. Lifelogging refers to the idea of recording your life and reviewing the images to better savor your existence.

I had the good fortune to be a part of a documentary film, *Lifeloggers*, that Memoto created, which features a number of great minds in the quantified self and wearable computing movements. It's also worth a watch to see how Csikszentmihalyi's ideas of experience sampling have evolved. To get a sense of how Memoto works, you can watch their Moment View Video, where pictures taken every thirty seconds are assembled into a slide show format.[6]

Lifelogging is not just a trend for the techie crowd. Recent studies have shown that reviewing photographs from your day can help in memory retention or even curbing dementia. Thought leader Brenda Milner from McGill University describes some of these types of findings in the video "Inside the Psychologist's Studio"[7] from the Association for Psychological Science.

Lifelogging can also provide a forum for activism, where a person's recorded photos or videos can be used for the common good. Steve Mann, whom many credit as being the father of wearable computing, coined the term "sousveillance" in a paper he wrote with colleagues Jason Nolan and Barry Wellman in 2003,

"Sousveillance: Inventing and Using Wearable Computing Devices for Data Collection in Surveillance Environments."[8] In the paper, Mann and his colleagues challenge a society that has become overly surveillance-focused to allow individuals to wear devices so they can record their own actions. Dubbing this activity "sousveillance," from the French words *sous* (below) and *veiller* (to watch), this practice lets people "track the trackers" in their lives. I interviewed Oskar Kalmaru, cofounder of Memoto, about lifelogging being used for this type of antisurveillance and other types of activism as well.

> Tracking yourself and the close environment around you has indeed proven to protect people from surveillance regimes. During the Arab Spring, an Egyptian blogger was accused of being involved in riots in Cairo, but thanks to having tracked his location, he could show that he hadn't even been in Egypt at the time. As self-tracking becomes even easier and more intertwined with the devices and services we already use, I think we will see an exploding number of use cases like these.[9]

In a blog post on the Memoto site, *Lifelogger* filmmaker Amanda Alm relates another, newer use of lifelogging that has become popular. Russian dashboard cameras have recorded footage of people being kind as well as unique documentation of a meteorite that crashed in 2013. In Alm's words, "Drivers having dash cams attached to their cars is common [in Russia], and what's captured can provide valuable evidence for situations like accidents, etc. Those dash cams were invaluable in gathering lots of material on that meteorite, something useful for scientific analysis."[10]

Lifelogging as empowered by the digital tools of quantified self is revolutionizing how we study flow. Like the Lively platform we discussed in Chapter Four, sensors are also helping us understand how our daily activities affect our health. A February 2013 report from the California HealthCare Foundation, *Making Sense of Sensors: How New Technologies Can Change Patient Care* by Jane Sarasohn-Kahn, opens with this fictional vignette describing how sensor technology could help monitor seniors for assisted living purposes:

> When Ann R., a sixty-five-year-old woman with congestive heart failure and diabetes, steps out of bed in the morning, her weight is recorded by a Wi-Fi–enabled sensor located

under the floorboards. This is just one of several sensors that keep a close watch on Ann's health without her having to do anything at all. The data detected by all the sensors are automatically transmitted via a secure wireless connection and stored in her personal health record in a cloud-based computer server. If any of the health measurement signals falls outside of a predetermined normal range for her, the data are transmitted to her clinician as well as to a family member designated by Ann.[11]

These types of sensors add a critical component to the study of flow that Csikszentmihalyi didn't have available for his initial work regarding the experience sampling method: passive data collection. While using a beeper throughout the day produced remarkable insights about people's behavior and well-being, their responses were still affected by survey bias. Meaning, on some level, they knew their answers were being recorded. They may also have been annoyed by the interruption of the beeper that could have affected their survey responses.

This doesn't minimize the early work done via the experience sampling method. On the contrary, the precedent established by Csikszentmihalyi's work means we can get the best of two complementary sets of data collected by active and passive means. As a reminder:

- Active data collection—people are aware their thoughts or actions are being recorded.
- Passive data collection—people give permission to be recorded, but then forget about the sensors or other tools tracking their behavior.

In Chapter One, I related three aspects of a person's well-being and identity in the Connected World:

- Subjective well-being—how you perceive your happiness and actions
- Avataristic well-being—how you project your happiness and actions
- Quantified well-being—how devices record your actions reflecting your happiness

It's our quantified well-being that is now adding a massive new dimension of insights to the data collected about our lives. The information generated about your actions can start to reveal what brings you meaning in ways we've never had access to before.

Flow in Action

In his *Psychology Today* article "Finding Flow," here's how Csikszentmihalyi talked about achieving a state of flow during leisure time:

> In comparison to work, people often lack a clear purpose when spending time at home with the family or alone. The popular assumption is that no skills are involved in enjoying free time, and that anybody can do it. Yet the evidence suggests the opposite: Free time is more difficult to enjoy than work. Apparently, our nervous system has evolved to attend to external signals, but has not had time to adapt to long periods without obstacles and dangers. Unless one learns how to use this time effectively, having leisure at one's disposal does not improve the quality of life.[12]

This is compelling information and runs contrary to the idea that we can get happier just by consuming as much entertaining media as possible. The article relates statistics saying that U.S. teenagers experience flow 13 percent of the time when they watch TV versus

34 percent doing hobbies and 44 percent playing sports or games. "Yet these same teenagers spend at least four times more of their free hours watching TV than doing hobbies or sports. Similar ratios are true for adults."[13]

The trick with flow is that a person needs to experience it regarding a new activity to understand how powerfully it can improve their lives. Discovering flow in your life requires investment that you'll only uncover if you track what works and what doesn't.

Sensor data is also starting to affect flow at work. Ben Waber, cofounder and CEO of Sociometrics Solutions and author of *People Analytics: How Social Sensing Technology Will Transform Business and What It Tells Us about the Future of Work*, wrote an article for *MIT Technology Review* about augmented social reality, an evolving technology platform that utilizes sensors to shape the physical environment of an office based on interpersonal relations. While flow as defined by Csikszentmihalyi involves a person understanding their behavior that provides optimal experience, this trend of data-supported social interactions will factor into how we create meaning in future work situations. Here's how Waber describes this new technology in "Augmenting Social Reality in the Workplace":

> Augmented social reality is about systems that change reality to meet the social needs of a group. For instance, what if office coffee machines moved around according to the social context? . . . By positioning [a] coffee robot in between two groups, for example, we could increase the likelihood that certain coworkers would bump into each other. Once we detected—using smart badges or some other sensor—that the right conversations were occurring between the right people, the robot could move on to another location. Vending machines, bowls of snacks—all could

migrate their way around the office on the basis of social data.[14]

I love visualizing how this might look in the future. Would these robots penalize me if I grabbed too many snacks? If I don't steam my espresso right, will the coffee machine ignore me for two weeks? In terms of helping to facilitate social interactions, I do see this technology providing worthwhile insights for companies willing to experiment. I'd also want employees to be able to access data from machine interaction to gain insights about their own behavior, however. In this way, employees could examine personal data that could give them hints about where they could find more flow at work.

Flow has been shown to increase with political activism. In an article for the *Pacific Standard*, "Get Politically Engaged, Get Happy?"[15] Lee Drutman reports on the work of two psychologists, Malte Klar and Tim Kasser, who found a link between political activism and a person's sense of well-being.[16] The pair were interested in studying eudaimonic happiness versus hedonic happiness with regard to civic engagement. Where life has a sense of purpose or direction, eudaimonia and flow can increase. Connecting with a group is also a form of happiness focusing on social well-being that Klar and Kasser were interested in studying regarding activism.

After conducting surveys with sample groups, asking about their histories and attitudes toward activism, the researchers discovered that being an activist correlated with being happy. Political activism gave people a greater sense of purpose and connection to community than those who weren't participating. We crave connection to our communities and to work for a purpose that is greater than ourselves.

Flow can also be experienced in educational settings. The Key Learning Community in Indianapolis has incorporated flow into their teaching methods to help encourage children to identify the areas where they find connection in their studies.[17] In an article from *Edutopia*, Csikszentmihalyi describes some of the ways the Indianapolis school has incorporated flow, including the use of a "Flow Room," where students can spend an hour a week exploring any subject that interests them. The school also hired a video technician who, along with teachers, interviewed each child at the beginning of the school year, asking them what they hoped to achieve by the end of the year. Throughout every semester, in a video-journal format, kids would update their tapes with recent accomplishments. Csikszentmihalyi describes the results:

> At the end of the year, the child could have a documentary of what he wanted to accomplish and what actually did happen. Now, to me—you know, you say, well, so what?—I think psychologically, it's a very important thing, because you are putting the responsibility for learning on the child. They are responsible for what they're going to learn.[18]

Kids are just like adults. We all crave autonomy with regard to pursuing what we love. Letting children understand the concept of flow means we're enabling kids to teach themselves.

The H(app)athon Project

My article in Mashable "The Value of a Happiness Economy" is what inspired this book. The article also inspired the nonprofit organization I founded called the H(app)athon Project. I have never experienced more creative flow than I did while working on these projects. I'll work on this book or H(app)athon starting at seven thirty in the morning and barely look up from my computer until around two p.m., when I realize I haven't eaten. So many of my core skills are utilized in these projects (researching, writing, networking, brainstorming) that I would do them for free.

I wanted to unpack our work with the H(app)athon Project as it combines a number of issues related to flow and positive psychology, along with sensors and happiness metrics. The Project is a reflection of the themes and premise of this book—that we're worth more than wealth.

I realize we haven't met (yet), but that's how I feel about you, by the way. I think you have inherent value, and the goal of creating the H(app)athon Project was to try to provide free tools to help people discover their value and then be encouraged to help others do the same.

The main tool we're providing is a free app that gets to know you via an interactive survey you take on your phone using active and passive data. More details follow, and you can go to www.happathon .com right now, in a spirit of action, to start learning what makes you tick and how you can get happier.

Basically, we're hoping we can help match what brings you flow to an opportunity to increase your personal well-being or volunteer to help others. We're "connecting happiness to action, one phone, one heart, and one city at a time" and you're welcome to join us in our work. After you read these details, you can learn more about the power of altruism in the next chapter, a powerful tool we're using for leverage for the H(app)athon Project to change the world for good.

Here's information we've been using to tell people about the H(app)athon Project:

The Problem

The GDP has led the world to focus on monetary wealth as the primary driver of happiness. People judge themselves or others based on the narrow lens of their worth defined by money. Human dignity is lost when we're forced to focus on stockpiling/consuming for ourselves at the expense of others.

How Can This Problem Be Addressed?

The science of happiness shows that long-term, intrinsic happiness (well-being) is increased by engaging in an action that inspires flow (optimal experience) and helping others (altruism).

How Will H(app)athon Help?

We are "connecting happiness to action, one phone, one heart, and one city at a time" to increase people's well-being in an atmosphere of transparent civic engagement. Our free H(app)athon app (available on any smartphone) will help people find dignity by recognizing their strengths and connecting them to action that increases their happiness and changes the world for good.

What We're Building and How

We've created a free app that can measure a person's well-being and actions as reflected by answers to our survey and the sensors in their smartphone. The personal happiness indicator (PHI) score that results is a reflection of their identity as quantified by data. By revealing a picture of who they are not based solely on wealth, we believe we'll increase their happiness.

The Next Step

Beyond simply revealing a person's PHI score, we're also going to provide suggestions for people to do good in order to increase their happiness. Science shows that people's happiness increases when they're given chances to help others. What we're providing in the mix is the unique new use of sensor data in mobile phones to identify a person's identity so as to better match people to volunteer opportunities that will increase their happiness. Nonprofits or other organizations get optimized, energized volunteers pre-screened for work that matches their needs.

Our Workshops

We've also created free workshops any organization can do that teach about the issues of economics, the science of happiness, and emerging technology. We have already done dozens of workshops around the world.

Our Traction

We've been featured in *USA Today*, the *Guardian*, *Forbes* (three times), *Fast Company*, BBC News, and Mashable. Our advisory committee has more than thirty people from organizations like the United Nations, the World Economic Forum, MIT, Salesforce.com, the University of Cambridge, and University of Pennsylvania, along with dozens of other experts.

Our Goals

The anonymized data from the surveys we have on our site now will form the predictive algorithms we're putting into our app. We're creating the matching criteria for people's PHI scores so they'll have opportunities for action and happiness after they use our app. We're planning dozens of workshops and have scores of videos on our site. We're giving people a free tool to recognize their individualized awesomeness not focused on money so they can change the world and Get H(app)y at the same time.

16

ALTRUISM

The best way to convince a skeptic that you are trustworthy and generous is to be trustworthy and generous.[1]

STEVEN PINKER

FOCUS ONLY ON YOURSELF or help others. Two choices, a ton of motives.

Altruism is a tricky concept. In the moment you're helping someone else, you're likely not thinking "I'm being altruistic right now" but "I'd better keep that toddler from walking into the street before she gets hurt." There are numerous reasons we may be compassionate or empathetic. Evidence shows that genetic makeup and learned behavior can also influence one's propensity to be altruistic.[2]

But let's be clear: Helping others means you also help yourself. There are physiological benefits for individuals when they're compassionate. There are sustainable monetary benefits for organizations utilizing shared value. Countries measuring citizens based on happiness indicators get a deeper, quantitative view of their citizens than they would if they measured only GDP.

Helping just yourself means you benefit others primarily through transactional means. You buy things and help a local store or economy. You're pleasant to others if it advances your needs. Are you evil? Not at all. Are you invested in others? Not at all. Does that affect your reputation? Yes, it does.

Psychologists use a term called prosocial behavior. This includes actions that benefit others outside of the intentions of people performing the actions. Altruism is prosocial but is also characterized by the selfless nature of behavior. Here's how the Altruistic Personality and Prosocial Behavior Institute clarifies this distinction:

> For the purpose of our study, we prefer a definition that relies on objective, measurable criteria. We characterize a behavior as altruistic when:
>
> 1. it is directed toward helping another;
> 2. it involves a high risk or sacrifice to the actor;
> 3. it is accomplished by no external reward;
> 4. it is voluntary.

Let's also be clear that compassion takes risk, and not just that people may think you're a softie. The real trial comes when you reflect on others' needs and feel their pain. It's hard to experience suffering. But we're built to help others and we have the capacity to improve personal altruism through practice.

If you're taking measure of your own life, it will benefit you to experiment with altruistic actions to see how it affects your health and well-being. On a global scale, leveraging our positive actions is a renewable resource that is scientifically proven to help both the giver and the receiver. In the case of flow, you have to participate in an activity to experience its benefits. Altruism and compassion function in similar ways. So why not get hooked on happiness?

Good Intentions

Forty percent of what makes us happy is based on our behavior. Sonja Lyubomirsky, a thought leader in the positive psychology movement, notes this finding in her book, *The How of Happiness: A New Approach to Getting the Life You Want*:

> What makes up this 40 percent? Besides our genes and the situations that we confront, there is one critical thing left: our behavior. Thus the key to happiness lies *not* in changing our genetic makeup (which is impossible) and *not* in changing our circumstances (i.e., seeking wealth or attractiveness or better colleagues, which is usually impractical), but in our daily intentional activities.[3]

Lyubomirsky has worked with thousands of men and women, observing behaviors of happy people to determine what actions drive their outlooks. One of her better-known studies focuses on a series of "happiness interventions" she conducted with two sets of participants. Both groups of people were asked to commit acts of kindness throughout the week (donating blood, feeding a stranger's parking meter) and keep "kindness reports" documenting their actions.

The first group was instructed to perform these acts throughout the week, while the second group did them all in one day. Both groups experienced a significant elevation in their happiness, although the people performing all their acts of kindness in one day had a larger increase. While this means the timing or regularity of committing these acts produced varied results, Lyubomirsky notes that "our study was the first to show that a strategy to increase kind behaviors is an effective way to elevate happiness."[4]

It makes sense that we get used to an act of kindness and it may lose its luster. This involves the idea of habituating a behavior in positive psychology, but this can be easily overcome. Walk a different route to work to meet new people, or go to the Random Acts of Kindness website to get inspired with specific ideas you can emulate, like this one:

GAS STATION

I was standing in line this morning at the gas station and there was a young mother with her child attempting to buy gas. She ran her debit card for five dollars and it was declined. She tried again to run her card for three dollars, but it was declined again. She left. I stood there heartbroken for this young lady. I didn't know her or her situation, but it touched me. I went to the door, and she was putting her child in the car seat. I told her to get ten dollars in gas and I would take care of it. I am thankful and fortunate to be able to do this small deed.[5]

Little deeds add up. And the fact that you feel better after doing them doesn't mean you were being selfish.

The Opportunity for Altruism

John Helliwell is professor emeritus at the Vancouver School of Economics at the University of British Columbia and was coauthor of the United Nations' first *World Happiness Report*. I had the opportunity to interview Helliwell about his research regarding happiness and social networks as well as some of his ideas on altruism.

He told me about recent research he'd been doing regarding happiness and inequality that revolved around gambling. As happiness is affected based on whether we're alone, in a group of strangers, or with friends, Helliwell and his team set up experiments where people were gambling alone or with partners to gauge levels of happiness. There were a number of interesting results based on when people experienced happiness for other people and their winnings as compared to their own gambling successes.

The most powerful insight, however, from the study came when people were given an opportunity for altruism. During the gambling experiments, John placed a table in the corner of the room with a basket for donations. A small sign noted that the people conducting the study were collecting money for a charity that would buy antimalaria nets to save lives. The goal was to see if people would share their gambling winnings if given the opportunity. Here's how Helliwell describes the results of this experiment:

> There was no pressure with this request for a donation. We just provided the opportunity to give. We measured happiness at all stages of this experiment and found a really big boost came from people giving money away. They got way happier, in fact, than the people who didn't give any money away. In fact, the happiness gain for people who were alone in the experiment was just as great as [it was] for people doing it in public view. It's not about what you get, it's about what you give. The biggest favor you can do for someone is give them the opportunity to do something generous.[6]

It's amazing that people gave away some of their winnings whether or not they were being observed, but also that their happiness measurably increased with the act. As they were alone, one would assume giving money away was a selfless act without the opportunity for selfish gain. And they still got happier as a result.

The Happy Hero

Heroes make a career out of being generous. Altruism is part of the job description. We've all wanted to be heroes, and altruism gives us the gift of leveraging these hidden intentions.

Dana Klisanin is a psychologist specializing in the use of arts and media to promote altruism and compassion. She contributes regularly to the Digital Altruism blog she maintains for *Psychology Today*. In her article "The Cyber-Bully vs. the Cyber-Hero," she outlines the importance of giving children positive role models in contrast to the cyberbullies that have received so much press. That's why she's created an award-winning interactive game called the Cyberhero League. As Klisanin describes, the game helps children counter cyberbullying by providing them a digital and real-world format to engage in positive, altruistic behavior. I interviewed Dana to better understand how kids could use technology for empowerment:

I've read from a few of your interviews that you're concerned kids are suffering from a lack of empowerment in the modern world. Can you explain what you mean by that?
Kids today are saturated with media. They have access to more information than ever before and through it they are learning about complex global challenges, especially human-caused climate change and social inequality. Unfortunately, they have limited power to affect the world. I am concerned that this lack of empowerment may lead to feelings of helplessness, apathy, and depression that may continue into adulthood.

Can you define cyber-altruism?
Digital altruism, or cyber-altruism, is altruism mediated by the Internet or mobile technologies. It requires the willingness to help another, access to a computer or smartphone, and a bit of our time, depending upon what the action involves. For example, clicking a link to donate food, water, or medicine doesn't take as much time as playing a computer game like Foldit, in which you contribute to scientific research by helping scientists learn more about folding proteins.

How can kids/people experience the benefits of altruism without being face-to-face in real life with someone else?

We don't need to be face-to-face with people in the real world to experience benefits—we can Skype with a friend overseas and enjoy it. We can play a massive multiplayer online game with a stranger in another country and find it enjoyable. Likewise, when we engage in digital altruism—when we take an online action that benefits someone else—we benefit as well. Altruistic action creates a ripple effect—goodness online impacts real people in the real world just as much as hateful actions do. As a society we've focused a lot of our energy on cyberbullying, for example, without teaching our children that there are alternate positive behaviors.

Do you see the Cyberhero League increasing compassion as well as altruism? Are they different?

Yes, the Cyberhero League is designed to promote a number of character strengths and virtues, including compassion. The cyberhero is a new incarnation of the hero archetype arising from the fusion of moral action and interactive technologies. The Cyberhero League is designed to promote this new archetype. To support our goal of increasing character strengths and virtues we have partnered with VIA Institute on Character, a nonprofit organization dedicated to advancing both the science and the practice of character.

What's the dream for the game?

The Cyberhero League is designed to support collaborative heroism. My dream is that the Cyberhero League will become a powerful force for tackling global challenges through extending the heroic journey across cyberspace. As a meta-level game it is a venue through which people of all ages can use interactive technologies act to act on behalf of other people, animals, and the environment. I dream that one day there will be a "cyberhero feature" in all interactive media—that the Cyberhero League icon will be integrated into interactive media and become as ubiquitous as those of Facebook, Twitter, Instagram, and Pinterest—facilitating a worldwide renaissance of human values and promoting the emergence of planetary consciousness.[7]

It's not enough to encourage kids to act heroically. We need to provide them models that show them how to do it. And without methodologies like the Cyberhero League, the benefits of altruism can't be introduced into the digital arena where kids can see its value to practice in the real world.

Compassion for Couples

Giving kids an opportunity to be heroic is a huge gift. A gaming environment gives them permission to be compassionate and see how others will react. For adults, it's also beneficial to empathize with others as a way to increase compassion. In a sense, we can gamify our experiences by pretending we're someone else to see how they experience life. A good example of this idea comes in Gary Chapman's Five Love Languages methodology. Chapman's idea is that people have five primary ways they feel most loved. If we know our partner's "language," we have a much better chance of being compassionate and communicating well. The languages include:

- Words of affirmation
- Acts of service
- Receiving gifts
- Quality time
- Physical touch

It's amazing how vividly you get to experience your character when involved in a deep relationship. Even when trying to be compassionate, missing signals from a loved one means you'll likely end up frustrated in many conversations. In my case, my wife and I learned about the Love Languages concept and it's helped us a great deal. I tend to be a "quality time" type, where Stacy is big on "acts of service." So when I was about to whine about watching a movie together a number of years ago, I decided to clean the kitchen instead. Then I vacuumed without being held at gunpoint. I didn't announce I was doing these things, but wanted to see if my acts would help our relationship. Guess what—in the moment, it sucked. It's housework. Nobody likes doing it. And then, of course, it hit me:

It's housework. *Nobody* likes doing it. Including Stacy.

I had always helped around the house before, but empathizing with my wife while doing the chores made me realize that I was being a big schmuck by asking her to spend quality time with me if it meant there was still housework to do afterward and I wasn't doing my part.

Curiosity can lead to compassion. Empathizing with someone else's interests is a great way to engage in altruistic behavior without it seeming like a chore.

Compassion Is Contagious

Research in 2008 from a study involving over 4,700 people who were followed over twenty years found that people who are happy increase the chances that someone they know will also become happy. Even more remarkable is the discovery that happiness can span a second degree of separation, increasing the mood of the second person's husband, wife, or close connection.[8] The study[9] was conducted by James H. Fowler and Nicholas A. Christakis and was based on detailed records originally collected for the Framingham Heart Study, conducted over twenty years, that studied a number of health issues, including smoking and obesity.

The study also documents how the influence of a social network could impact policy change as well as health improvements. In the article, for instance, Fowler notes that "whether a friend's friend is happy has more influence than a five thousand dollar raise" with regard to increased well-being for participants, positing a focus on happiness would be a better gauge of national health than the GDP.[10]

In terms of virtual currency, this flow of goodwill may constitute a new part of the happiness economy that's already been created with metrics like Bhutan's Gross National Happiness Index. Where people's upbeat moods can increase well-being two degrees away, social networks could literally be paying their appreciation for people's actions.

"Happiness *is* contagious." Nataly Kogan is the chief happiness officer and cofounder of Happier, a company whose app encourages people to take photographs of things that inspire them to share with friends. I interviewed her and asked her how acts of happiness can inspire compassion.

> It's pretty simple. Happiness comes from taking note of small positive things and sharing them. It's about things like smiling more, or saying hi to strangers. When I first started doing this stuff, I thought it would be a farce or have to be complicated. It turned out to be the opposite. Once I started keeping a gratitude journal and studying the science of happiness, I learned that people who are more positive are healthier and less depressed. People tend to chase the wrong things and end up missing what could already bring fulfillment in their lives.[11]

The Currency of Kindness

It's going to be more difficult to avoid being compassionate in our connected future. People may record you walking by a homeless person without a second glance. On the highway, cars connected with M2M (machine-to-machine) technology may register when you cut someone off. Actions that have been ignored in the past will now be recorded. While the threat of accountability-based influence may not be the best incentive to inspire altruism, it's a start. And once you begin experiencing the happiness associated with compassion, you won't want to stop.

THE VALUE OF
A HAPPINESS ECONOMY

You know what's truly weird about any financial crisis? WE MADE IT UP. Currency, money, finance, they're all social inventions. When the sun comes up in the morning it's shining on the same physical landscape, all the atoms are in place.[1]

BRUCE STERLING

THE IDEA OF THE economy as a concept has always been elusive to me. It always seemed boring. Much of economics is based on statistics, and I was never great with numbers. Analyzing data about global populations has always mystified me, and I've avoided thinking about economics because it seemed so foreign to my interests.

I've changed my perspective.

When you measure something, you analyze one specific data point—you step on a scale to determine your weight, for instance. But measurement also involves *intent*. Why are you weighing yourself? For an annual physical, where your weight could determine follow-up care, or to see if you'll fit into a bikini? You use the same scale for both procedures, but for very different ends.

Currency was invented by humans. Squirrels don't use nuts to buy dental floss. As noted in Chapter Ten, a monarch from days gone by minted coins with his image and mandated they be used as a form of exchange. This forced citizens to provide food and clothes to soldiers, and eventually they adopted the coins as being a symbolic representation of stuff.

I found a great definition and description of economics on a website called Investopedia:

DEFINITION OF "ECONOMICS"

A social science that studies how individuals, governments, firms, and nations make choices on allocating scarce resources to satisfy their unlimited wants.

INVESTOPEDIA EXPLAINS "ECONOMICS"

Classical economists believe that markets function very well, will quickly react to any changes in equilibrium, and that a "laissez-faire" government policy works best. On the other hand, Keynesian economists believe that markets react very slowly to changes in equilibrium (especial[ly] to changes in prices) and that active government intervention is sometimes the best method to get the economy back into equilibrium.[2]

Economics may be based on statistics, but the analysis is based on the interpretation of economists. Economists have intent and bias when they report on the present and future state of the world.

My bias is that you should own your data, as it exists now and in the future. This is key to your future happiness. My intent is to demonstrate that the exchange of data in our current Internet economy is an occluded and confusing process. It favors data collectors over data producers (you) in an unfair value exchange.

Here are some equations to reiterate my point:

Our current Internet economy:

    We use computers a lot +

    Companies track our behavior while we're on computers +

    This tracking is interpreted as data that represents a form of our identities +

    Most of us don't understand how much we're exchanging for "free" services +

    Data brokers are not legally mandated to reveal the data they've collected about us

    = *Data brokers sell our data/identities for money.*

    = *We don't get paid in this process.*

    = *Transactional value is not transparent in this process.*

    = *This process has eroded trust.*

Our Internet economy in the near future:

    (Nothing changes from the previous equation) +

    Augmented reality expands data collection into the unchartered virtual realm +

    Data from sensors never available before broadcasts our personal information +

    Machines and objects around us collect and broadcast data about us at all times

    = *Our identities and actions are exposed and for sale without our full knowledge or control.*

    = *There are no set standards for all of this data collection/ exposure.*

    = *There are no ethical standards for all of this data collection/ exposure.*

The key issue in these equations is one of economics, not privacy. And as Investopedia defines it, economics is a *social* science. These issues of data are social in nature.

So here's my bias for this book: You're worth more than money. Your data and identity are not just items to be sold; they represent who you are. Data is currency, the same as if it were minted by a monarch, the same as if *your* image were on a coin.

Your data is you. If you don't own or recognize it, how can you measure its value for your life?

So here's a new equation for today and the future:

The happiness economy:

People own and control their own data +

Data quantified about our lives is oriented to reveal actions and words that give us meaning +

We take measure of our lives in ways we never have before +

We improve our happiness and well-being by actions informed from self-analysis +

We utilize happiness indicator metrics for a new global standard of value measurement

= *We base our well-being on transparent and ethical metrics.*

= *We have positively oriented personal and global measures of value.*

This vision is not metaphorical. The technology to measure and evaluate a happiness economy exists today. We're at a seminal crossroads in history that does not have to lead to an inevitable technological dystopia.

- We can break open the data model that exists today and leverage the information about our lives to increase well-being and positive change.
- We can stop thinking of ourselves as broken vessels in need of repair versus individuals ripe with promise, poised for greatness.
- We can embrace self-measure and empowered action to change our lives for good.

We can Hack H(app)iness.

The Value of a Happiness Economy

The Mashable article I wrote that inspired this book provides a number of examples of how the happiness economy already exists in a number of environments today that haven't coalesced under one formal taxonomy.

Yet.

I've reprinted the article as a way to set the stage for future chapters about happiness metrics trending around the world to prove that a happiness economy is as viable and real as the current economy we've created for ourselves.

THE VALUE OF A HAPPINESS ECONOMY

What if generosity were a currency? This was a question posed by the Danish chocolate company Anthon Berg for its recent Generous Store campaign. The company opened a pop-up store for one day in Copenhagen last winter and distributed chocolate as payment to individuals who promised to perform a generous deed for a loved one.

Chocolate lovers posted to the company's Facebook page, sharing promises like "serving breakfast in bed." Then they picked up their chocolate payment at the store and essentially broadcast to their social graph to "pay it forward."

Research suggests that paying it forward is something the average person enjoys. Søren Christensen, a partner in Anthon Berg's ad agency, says his company's findings showed that seven out of ten people were happy when they did something good for other people. But only one out of ten people ever experienced generosity on a daily basis.

Why the disparity, and why does it matter?

Two reasons. First, there's a growing movement to standardize the metrics around well-being that can lead to happiness. Second, the combination of Big Data, your social graph, and artificial intelligence means everyone will soon be able to measure individual progress toward well-being, set against the backdrop of all humanity's pursuit to do the same. In the near future, our virtual identity will be easily visible by emerging technology like Google's Project Glass and our actions will be just as trackable as our influence. We have two choices in this virtual arena: Work to increase the well-being of others and the world, or create a hierarchy of influence based largely on popularity.

Metrics, Not Mood

If you're thinking the study of happiness and well-being seems flaky, you're missing a major trend that's beginning to influence a number of global economies.

At the recent United Nations Summit, Secretary-General Ban Ki-moon stated that "Gross National Product (GDP) fails to take into account the social and environmental costs of so-called progress." In other words, measuring well-being is not the pursuit of identifying the ephemeral emotion of happiness. It's about looking at a deeper level of "economic, social, and environmental objectives that are most effectively pursued in a holistic manner."

And economics alone are not the primary driver of well-being. Statistics show, for instance, that after a person or family receives a salary of $75,000 per year, increasing the amount of money brought home doesn't increase a feeling of well-being.

Jeffrey Sachs, the renowned economist from Columbia University who edited the first *World Happiness Report* for the UN, certainly comes to the same conclusion. He said, "The U.S. has had a three-time increase of GNP per capita since 1960, but the happiness needle hasn't budged." The report, which provides scientific evidence that happiness can be reliably measured and is meaningful, notes that the U.S. [is] not as happy as other countries because of a too-prominent focus on boosting the economy—while largely ignoring long-term effects on environment or holistic education. (The Danes, however, were listed as the happiest people on the planet by Sachs's report—apparently Anthon Berg is onto something with their Wonka-onian economics.)

H(app)athon

The study of happiness is a burgeoning field of study around the world, with scientists and other experts providing hard data as to the benefits of a balanced approach to well-being versus too singular a focus on money or self.

"Our goal is to get people thinking more deeply about what happiness is and what is the connection between themselves and their community and world," says Laura Musikanski, the executive director and cofounder of the Happiness Initiative, an organization inspired by Bhutan's ideas on Gross National Happiness, also known as GNH. They even created a survey geared to measure ten metrics of well-being, which include material well-being, physical health, and time balance.

Her site also contains an excellent history of happiness research that provides important data-related insight. For example, although ephemeral happiness may come about due to a combination of luck, timing, or fate, the emerging science of happiness proposes that "our actions determine 40 percent of happiness, and that well-being can be both synthetically created and habitually formed."

This may be the biggest reason for our desire to measure this space, and several takes on measuring it have popped up. The quantified self movement has exploded, and Nicholas Fenton's practice of chronicling information for his annual life's report has inspired others to follow his lead via Daytum and other self-monitoring services. Ariana Huffington also recently announced her GPS for the Soul, an app that provides a "course-correcting mechanism for your mind, body, and spirit."

The natural next step in this process, then, is to marry the collective metrics of individuals to form a collective virtual picture of a community or country. Mirroring the goals of GPS for the Soul, it would be simple to map GNH/well-being metrics to existing technology like Mint.com that provides updates on how to maintain material well-being or Project Noah that encourages more access to nature. Via this methodology, our lives could become a virtual h(app)-athon, with technology doling out advice on how to flourish, while proactively helping others.

The Efficacy of Fun

But as with any behavior or state of mind, it will take a village. "A really important part of changing behavior is social reinforcement," says C. Lincoln (Link) Hoewing, assistant vice president for Internet and Technology Issues for Verizon and frequent contributor to Verizon's Policy Blog. "You start seeing and comparing yourself to others more when you know that other people can find out what you're doing."

This form of accountability-based influence (ABI) is most effective when eliciting a positive response. As an example, Hoewing noted Volkswagen's Fun Theory campaign, whose Piano Stairs YouTube video has received almost 18 million views to date. For the campaign, a set of stairs in a Stockholm subway [was] outfitted with full-size piano keys that played notes as people walked on them, resulting in 66 percent more people than normal choosing the stairs over the nearby escalator. It's a simple leap to picture this event being geared toward a community metric of well-being, where the GNH for Stockholm would have risen the day of the campaign.

The Currency of Community

Brands are certainly learning to leverage well-being in the form of corporate social responsibility known as shared value. While bringing happiness to consumers via a product or service is not unique, bringing happiness to a community is just coming into widespread acceptance. "We want to set in motion an upward spiral of confidence," stated Starbucks CEO Howard Schultz in his Letter to America last August. This included the company's Create Jobs for USA program, which has seeded $5 million to provide capital grants for un-derserved community businesses.

"The idea of the initiative is to create happiness coming from economic well-being," states Adam Brotman, chief digital officer for Starbucks. The company also recently announced its Store Partnership Model, where pilot community organizations in New York City's Harlem neighborhood and Los Angeles's Crenshaw neighborhood will share in the profits of a Starbucks store. A minimum of $100,000 for each organization will seed programs geared toward job and life skill development, positive learning environments, and overall health and wellness in the community.

"We're in the happiness and people business," says Brotman, referring to the shared value mentality that a social business can be generous and profitable at the same time. "A thriving or happy community is something that's good for everybody."

When Actions Create Identity

In about three to five years, it won't matter if you'd rather not project your actions to the world—your virtual footprint will simply be too hard to conceal. Your preferences combined with the data generated by external forces will in essence make everything, including objects, inherently interactive.

"What's a social network for data?" asks Jim Karkanias, an executive at Microsoft who runs the company's Health Solutions Group, and has been working on a range of projects that broach the physical and computing worlds. "We're imagining biology versus silicon as the next platform in which we write software." Karkanias uses a form of prototyping for his work based on Project Hieroglyph, a movement that encourages science fiction writers to infuse their work with optimism that can inspire a new generation to "get big stuff done." "Science fiction sets the stage for people to imagine things bigger than reality," says Karkanias, noting that adhering to practicality in ideation tends to create a narrow experience that limits imagination and hinders happiness.

Data already has its own social networks: RFID tags, M2M (machine-to-machine) sensors in cars, and the Internet of Things let machines trade information without the need for human intervention. The self-tracking craze with humans combined with this ubiquitous data means highly personalized and proactive information can be aggregated to inform our actions on a minute scale. The advent of things like Google Glass means we'll be able to virtually see other people's data as well as eventually record our entire existence. Our lives will be tagged and ranked as semantic information fed into a massive global algorithm that could be geared toward inspiring positive behavior.

Karkanias agrees: "Artificial intelligence in the form of a perpetual life coach will live at the information level providing guidance on every aspect of your day."

Technology of this kind will likely manifest itself in a reverse Siri interface, with a GPS-like voice guiding you on issues both personal and macro. The societal impact could shift negative personal patterns as well as a community or country's Gross National Happiness.

Karkanias provides an example of this model where you're in your car and take a route that passes a McDonald's. As your coach knows your health issues regarding cholesterol, it adjusts the route of your self-driving car to the nearest Whole Foods to map to your GNH/well-being metric regarding health. Likewise, cameras in a subway car utilizing facial recognition technology might scan the face of a woman who is four months pregnant and send you a text to give her your seat to map to her GNH/well-being metric of psychological well-being. Emerging services like Sickweather will provide health-related predictive data that will affect whole communities regarding metrics of time, balance, and well-being.

Inspiration versus Ignorance

Some pundits say that privacy is disappearing, but that doesn't mean we should let our identities be dictated by outside forces. Unfortunately, people are largely unaware of the repercussions of giving away personal information as we enter a virtual era where information can be accessed by so many parties so easily.

"People are not fully aware of the data they generate and how that's coupled with artificial intelligence learning algorithms. It's creating a different social and economic order, and we're in the midst of that happening now," states John Clippinger, founder and executive director of idcubed.org and a scientist at the MIT Media Lab Human Dynamics Group, where he is conducting research on trust frameworks for protecting and sharing personal information. He feels the inevitable onset of ubiquitous data meshing with synthetic biology and people's social graphs can be a positive evolution if the whole process takes place in the open.

This transparency is the key. Fostering a culture based on GNH and mapped by existing technology provides a positive path toward the future. We should emulate chocolatier Anthon Berg and let generosity be our currency. Our lives will be sweeter for the choice.[3]

BEYOND GDP

Too much and for too long, we seem to have surrendered personal excellence and community values in the mere accumulation of material things.[1]

ROBERT F. KENNEDY

LOOKING INWARD IS *HARD*.

Helping other people is *hard*.

That's why we don't do these things too often. We have the perfect excuse: We don't have the time, because we need to earn more money. Yet there's no set number for the perfect amount of money to earn, no set definition of well-being that comes from a certain amount of wealth. We've just been told it's essential to build the economy by producing and consuming as much as possible.

Gross domestic product as a *measurement* is a useful tool; it's a standardized way of looking at the world everyone has agreed on for more than a half century. Gross domestic product as a *philosophy* is killing us. A focus on quantity over quality means you aren't ever asked to look inward, unless it's to dig deeper to create more wealth. A focus on quantity over quality means you aren't ever asked to help others, unless it provides a tax break or a reputation increase.

We have been forced to surrender our personal excellence. We don't have time to fully reflect on how we could uniquely contribute value to the world. We have been forced to surrender community values. We don't have time to help others if it will diminish our opportunities to produce more wealth.

Surrender. Yield. Consume for the sake of consuming. All you are worth is what you are worth.

No.

Screw you and the horse you rode in on. I'm done surrendering. It's time to reimagine how the world sees wealth.

Beyond GDP is a movement that's not actually based on measures of happiness. Well-being or metrics around quality of life are often involved, but Beyond GDP refers to work done in the past thirty years around the world by countries and organizations looking to find a new measure of economic wealth versus gross domestic product.

Robert Kennedy

In a famous speech delivered at the University of Kansas in 1968, Robert Kennedy outlined why gross domestic product was such a harmful measure of value. Credited as the beginning of the Beyond GDP movement, he outlined in his speech how the GDP prioritizes negative measures while ignoring other essential areas altogether:

> Even if we act to erase material poverty, there is another greater task. It is to confront the poverty of satisfaction— purpose and dignity—that afflicts us all . . . Our gross national product, now, is over $800 billion dollars a year, but that gross national product—if we judge the United States of America by that—that Gross National Product counts air

pollution and cigarette advertising, and ambulances to clear
our highways of carnage. It counts special locks for our
doors and the jails for the people who break them. It counts
the destruction of the redwood and the loss of our natural
wonder in chaotic sprawl . . . Yet the gross national product
does not allow for the health of our children, the quality of
their education, or the joy of their play . . . It measures neither our wit nor our courage, neither our wisdom nor our
learning, neither our compassion nor our devotion to our
country. It measures everything, in short, except that which
makes life worthwhile. And it can tell us everything about
America except why we are proud that we are Americans.

If this is true here at home, so it is true elsewhere in the
world.[2]

Confronting the "poverty of satisfaction" is a difficult challenge.
It means we're forced to look beyond money to see what's truly
worth measuring in our lives. As Kennedy points out, the GDP
credits things like "the destruction of the redwood." While measuring the erosion of the environment is necessary, a dangerous
precedent has been set with the creation of "offsets" regarding
negative issues like pollution. Rather than work to eradicate the
spread of dangerous environmental practices, countries are permitted to continue what they're doing if incentivized to replace
what they've destroyed. But these offsets are simply a delaying
tactic for the inevitable when dealing with finite resources. It's a
justification of destructive and irreversible economic practices
that embodies the poverty of satisfaction.

The landscape is changing, however. Sensor technologies and
quantified self tools are helping people measure the quality of
their health on a wide-scale basis. Positive psychology measures
how our compassion can increase our happiness while putting others above our constant quest to consume. We are at a crossroads of

time where we can measure the things that make life worthwhile if we supplant our reliance on money as a primary means of value.

Gross National Happiness

It was four years after Kennedy's speech that Bhutan's fourth Dragon King, Jigme Singye Wangchuck, coined the term Gross National Happiness (GNH). Spoken as a casual remark, Wangchuck felt the GDP did not serve as an accurate measure of value for his country based on Buddhist spiritual values. The term resonated with Wangchuck's colleague, Karma Ura, who created the Centre for Bhutan Studies along with a survey tool that measured Bhutan's well-being via a methodology quite different from the GDP.

While the name may imply a focus only on increase in mood, Gross National Happiness actually comprises measure of the four following pillars. And, as Wikipedia points out, "although the GNH framework reflects its Buddhist origins, it is solidly based upon the empirical research literature of happiness, positive psychology, and well-being."[3]

- Promotion of sustainable development
- Preservation of cultural values
- Conservation of the natural environment
- Establishment of good governance

These pillars were later refined to include the following eight contributors to happiness:

- Physical, spiritual, and mental health
- Time-balance
- Social and community vitality
- Cultural vitality
- Education

- Living standards
- Good governance
- Ecological vitality

You'll note that living standards are a contributor to happiness—money always plays a role in determining someone's well-being, but isn't the only contributor to happiness. The Bhutan model stresses that these contributing factors need to be in balance before a person can be in a place to pursue happiness.

If your time balance is out of whack and all you do is work, it won't make a difference how much money you have regarding your well-being. If you are in good physical shape but don't have access to educational resources, your happiness will also be affected. The number and types of indicators in GNH have been challenged since it first came into being. But the idea that they provide a better overall reflection of a country's value than just monetary measures is a primary reason GNH has driven awareness for the Beyond GDP movement overall.

The What and the How

Three quick points about measuring well-being that are important to note while studying the evolution of the GDP Movement.

First is a concept known as Maslow's hierarchy of needs. Proposed by psychologist Abraham Maslow in 1943 in his paper "A Theory of Human Motivation,"[4] the concept posits that humans have basic needs they need met before they can focus on deeper levels of intrinsic fulfillment. This is why postulating on theories of happiness to a person without potable drinking water doesn't make much sense. If someone is primarily focused on attaining basic needs to survive, achieving happiness for its own sake can be very difficult.

Second, it's important to note *how* metrics like Gross National

Happiness are measured. Along with census reports or other common public data, statisticians or social scientists utilize surveys to ask citizens to self-report their own levels of life satisfaction. Along with survey bias, the issue of participants answering questions based on knowing how their responses will be measured, surveys also are created with intent. This doesn't infer they are nefarious in nature, only that how questions are posed and arranged in a survey can directly affect responses.

Finally, most of us tend to think that large-scale survey results are a form of quantified data when technically they're actually the aggregation of multiple subjective answers. This is an important nuance to note: It means that policy or other decisions are being made on the collective and potentially biased responses of participants for any survey.

This is not to disparage data taken from surveys, but to note how the science of behavior will evolve in the near future. As mobile sensors become an accepted way to provide "answers" to surveys via passive data, the nature of subjectivity in responses will change. The Nobel Prize–winning psychologist Daniel Kahneman created a methodology called the Day Reconstruction Method[5] as a measurement in social science where participants record their memories from the previous day in response to a survey or experiment. By definition people's responses are subjective (their own truths) and also suffer from human error—we don't always remember even recent facts about our lives with accuracy.

In a future incorporating mobile sensors, our activities will operate like our credit card bills: We'll receive reports based on what we've actually done versus what we've remembered. For surveys based on happiness and well-being metrics, these reports may also take some getting used to.

For instance, my friend Neal Lathia, a senior research associate in the Networks and Operating Systems Group of Cambridge University's Computer Laboratory, has created an app, called

Emotion Sense, that collects data from all the passive sensors that a phone provides, including ambient noise. The app utilizes surveys about emotions and satisfaction with life that have been defined by psychologists to seek new insights previously unavailable, as mobile phones couldn't collect this data before passive sensors existed. While a person may remember their previous day as being positive, Emotion Sense might have logged multiple times where a person's voice registered frustration. While any app, like a survey, is influenced by the intention of the people who created it (and Neal has coauthored a paper on this issue, "Contextual Dissonance: Design Bias in Sensor-Based Experience Sampling Methods"),[6] these types of sampling methods will be a complementary objective measure in the survey world. For example:

•   If you're asked to recall physical activity for a survey, you might only register your exercise, where a pedometer could measure actual steps you took, even if they were to and from your refrigerator. The accelerometer sensor in a smartphone can also tell the difference between motions related to sitting, standing, or active movement.

•   You may rate a previous day as being negative largely based on the last thing that happened during the day. Daniel Kahneman, founder of behavioral economics, calls this phenomenon the riddle of experience versus memory, as our "experiencing selves" and our "remembering selves" perceive events in different ways.[7] For instance, if you have a root canal for two hours and experience steady pain for the first ninety minutes but the last thirty is comfortable in comparison, Kahneman's research shows many of us will report that session as not being painful. However, the reverse is also true: If the majority of the two hours is comfortable but you experience a great deal of pain in the final minutes of the session, you'll recall the entire event as painful. These types of relationships will be revealed more often in the future as sensors become prevalent for data collection in surveys.

• We may soon get to the point where information from outside sources regarding a survey could come into play regarding data collection. In the same way three reporters can cover a story three different ways, other people's responses to your actions may begin to factor into surveys you take. For instance, say you self-reported that a previous day had been fairly calm. If two other people using Google Glass devices recorded you at a Starbucks where you raised your voice at a barista, their impressions of your behavior might register differently than your own memory. Sensors measuring your behavior outside of the ones you're wearing will likely come into play very soon for large-scale measurement of happiness and well-being.

Why Happiness?

Our politicians, media, and economic commentators dutifully con-
tinue to trumpet the GDP figures as information of great portent . . .
There has been barely a stirring of curiosity regarding the premise
that underlies its gross statistical summation. Whether from sin-
cere conviction or from entrenched professional and financial
interests, politicians, economists, and the rest have not been eager
to see it changed. There is an urgent need for new indicators of pro-
gress, geared to the economy that actually exists.[8]

—CLIFFORD COBB, TED HALSTEAD, and JONATHAN ROWE

Jon Hall is the head of the National Human Development Reports Unit (HDRO), part of the United Nations Development Programme. Before joining the HDRO, Hall spent seven years working for the Organisation for Economic Co-operation and Development (OECD), where he led the Global Project on Measuring the Progress of Societies. He recently was lead author of Issues for a Global Human Development Agenda for HDRO and has been actively involved in research around human development and well-being since 2000. He is also advising the H(app)athon Project and provided me with a history of the Beyond GDP movement in an interview for *Hacking H(app)iness.* "Happiness is a single number that goes up and down and can easily be interpreted," Hall noted in our interview, pointing out the power of a single and simple unifying metric to measure well-being. "That's how I became converted about happiness as a single overarching metric of progress."[9]

As a statistician, Hall became aware early in his work around well-being how important it was to create metrics that the average citizen could understand. As he noted in his interview for this book, after developing a groundbreaking report for the Australian Bureau of Statistics called *Measuring Australia's Progress*, Hall showed the report to his mother for her thoughts. "She said, 'I love the picture of my grandson on the cover, but it has too many numbers, so I gave it to your father.'"

It's a fun story, but points out the disconnect Jon has been trying to eradicate for years: How can we make what we measure matter to the average citizen? One way is to keep things simple, which is why Jon favors the use of happiness metrics to measure life satisfaction.

Happiness provides a broad summary of how people feel about their lives. It doesn't require statisticians or economists to add things up and take an average. Adding metrics together like health or education often gets people upset because the formulas complicate things: How does an extra year of life compare to an extra year of education, for example? There is no "right" answer. Happiness relies on people answering a simple question about how they feel.[10]

One of the more popular ways to measure life satisfaction or happiness is called the Cantril Ladder Scale (or self-anchoring striving scale), developed by Hadley Cantril in 1965 at Princeton University. People rate their lives based on an imaginary ladder where the steps are numbered zero to ten. Zero represents the worst possible life, and ten represents the best.

While people's responses to these questions are inherently subjective, they do provide a way for people to simply express how they feel on a certain issue. As Hall points out, the more you collect these kinds of assessments, the more insights you also gain about a community, and the more the results can inspire conversations about things that matter.

> One aggregate number changing around happiness is easy to collect and it can fluctuate for a multitude of reasons based on people's concerns. The difference between socioeconomic groups can also be telling: If the measure goes up or down for a certain group, then that can trigger a conversation. Imagine if the government regularly reported measures of happiness, and imagine if happiness for white young women in Florida dropped last week, then that might trigger a broader debate in the media. Is it because of recent reports of rape? Is it because of reports that women in the state earn less than men? . . . And so on, and so on. The statistics can provide a window into a broader set of conversations that go beyond GDP: They can be very powerful ways to inspire much broader and informed debates around issues that are relevant to citizens.[11]

Social media is providing another platform for studying happiness and well-being that can help to soften our single-minded focus toward the GDP. The World Well-Being Project (WWBP), part of the Positive Psychology Center at the University of Pennsylvania, is leveraging the massive availability of citizen commentary available through social media platforms to analyze psychosocial and physical well-being. By measuring language through sentiment analysis and other similar methodologies, the WWBP also gets to leverage the scale of social media to analyze large data sets for mining insights around happiness and well-being. The group's vision is that their "insights and analyses can help policy makers to determine those policies that are not just in the best economic interest of the people, but those which indeed further people's well-being."[12]

One of WWBP's studies, "Characterizing Geographic Variation in Well-Being Using Tweets," showed that "language used in tweets from 1,300 different U.S. counties was found to be predictive of the subjective well-being of people living in those counties as measured by representative surveys."[13] A word cloud visualization of the report can be seen below. This practice of utilizing social media to study well-being sets a fascinating precedent, especially if results mirror surveys focused on the same issues. Citizens get to utilize transparent tools to broadcast their emotions, while organizations like the WWBP help analyze sentiment that could lead to policy change.

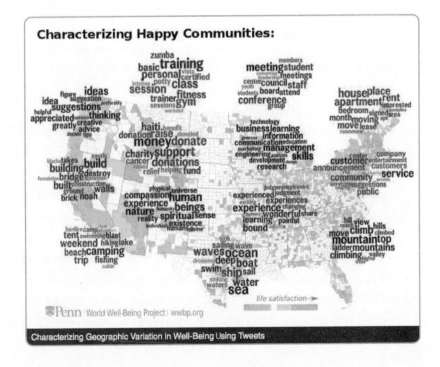

Characterizing Geographic Variation in Well-Being Using Tweets

Evolution

From 2000 to the present, a number of initiatives and organizations have helped push the envelope to move Beyond GDP. One big push came from the Organisation for Economic Co-operation and Development (OECD). In the early 2000s, the organization created a focus on measuring well-being for its thirty or so member states focused on global development. As Jon Hall noted in our interview, "The OECD getting involved in this work was a big stamp of approval; it gave the study of happiness and well-being a whole new level of seriousness."[14]

The organization created an interactive Better Life Index, which lets users rate eleven topics in real time to see how OECD member countries compare based on each issue. The topics reflect what the OECD feels are essential to both material living conditions (housing, income, jobs) and quality of life (community, education, environment, governance, health, life satisfaction, safety, and work-life balance).

In 2007, the OECD held a world forum on Measuring and Fostering the Progress of Societies in Istanbul, and the EU organized a Beyond GDP conference, hosted by the European Commission, the European Parliament, the Club of Rome, the OECD, and the World Wildlife Fund, which met to discuss which indicators would be most appropriate to measure progress in the world. In 2008, President Nicolas Sarkozy of France created the Commission on the Measurement of Economic Performance and Social Progress focused on evolving the GDP. The commission was chaired by renowned economist Joseph Stiglitz.

After a number of other initiatives took place over the following years, in 2012 the United Nations implemented Resolution 65/309, a provision that permitted the Kingdom of Bhutan to convene a high-level meeting as part of the sixty-sixth session of the UN General Assembly in New York City. The Resolution recognized "that the gross domestic product . . . does not adequately reflect the happiness and well-being of people," and "that the pursuit of happiness is a fundamental human goal."[15]

On April 2, 2012, the prime minister of Bhutan convened the summit titled, "Wellbeing and Happiness: Defining a New Economic Paradigm." The report from the meeting outlines a series of next steps attending organizations are taking to implement measures of well-being and happiness to move Beyond GDP. The prime minister of Bhutan, H.E. Mr. Jigmi Y. Thinley, opened the meeting with the following remarks:

> We desperately need an economy that serves and nurtures the well-being of all sentient beings on earth and the human happiness that comes from living life in harmony with the natural world, with our communities, and with our inner selves. We need an economy that will serve humanity, not enslave it. It must prevent the imminent reversal of civilization and flourish within the natural bounds of our planet while ensuring the sustainable, equitable, and meaningful use of precious resources.[16]

The meeting was a watershed event in terms of the world formally looking to dismantle or at least complement the GDP with factors directly related to increasing measures of well-being, happiness, and flourishing that cannot be created or sustained by money alone.

Building Genuine Wealth

Mark Anielski is president and CEO of Anielski Management in Edmonton, Alberta, and author of the best-selling book *The Economics of Happiness: Building Genuine Wealth*. I interviewed Anielski about his ideas based on his experience in Canada with natural capital accounting and work developing alternative measures of economic progress beyond the GDP. One of the initial aspects of his book I found so compelling was in the introduction where he states, "Economics is more like a religion than either art or science."[17] I asked him to elaborate on this idea.

> Economics is like a religion because it demands that society accept certain axioms, theories, principles, and suppositions about how human beings behave. Economics would have us believe that all people are consumers measured in terms of GDP, where everything is valued in terms of a money "price" that mediates all transactions and human relations. However, the idea that all people behave in a similar fashion in some hypothesized maximization of utility is a convenient simplification of how people actually behave. The trouble is, if you don't believe in these theories, then you find yourself outside of the "religion" of neoclassical economics. But the truth is, people are not rational and do not behave the same. There is no such thing as a perfect market.[18]

It's helpful to redefine economics as a study of people's behavior versus simply the output of their labor. Well-being is multifaceted and relies on numerous cultural biases and predispositions that can't be unilaterally measured by algorithms or indexes that don't truly represent how people or communities work.

According to Anielski, Genuine Wealth must inherently factor in the human and environmental capital of a country or it won't accurately measure what truly brings well-being to its citizens. And one of the reasons we've stayed with the GDP as a measure of value for so long is simply how hard it is to calculate these attributes as compared to utilizing fiscal metrics. But with Gross National Happiness and the Beyond GDP movement gaining traction, governments around the world are being pressured to use more transparent and accountable methods to measure what truly brings contentment in modern society. I asked Anielski about these ideas in relation to my thoughts on accountability-based influence and how our actions on individual and collective levels would begin to alter how the world views wealth moving forward:

> I believe that genuine accountability will result when businesses and governments operate from the basis of a true balance sheet that keeps an account of the physical and qualitative conditions of its human (people), social (relationships), and natural capital assets. In the Genuine Wealth model I've developed, the measures of progress and proxies for the resilience or flourishing of assets will be tied to virtues, values, and principles for what people intuitively feel contributes most to their well-being. We will then be able to confidently say we are measuring what matters.[19]

The Present and the Future

Implementing changes based on happiness will take more time. But whereas many leading economists ten years ago discounted the study of well-being and happiness as frivolous, as Jon Hall notes, "they've changed their minds." I asked Hall where he felt the Beyond GDP movement and happiness metrics would evolve in the future:

> In five years' time I think people will be using this type of data to implement policy. In twenty years this could be very radical. Well-being could actually change the way that the machinery of government is put together. We'd have a realignment of how different ministries work together and how decisions are made. It will change everything.[20]

I also had the pleasure of interviewing Enrico Giovannini for his thoughts on well-being and happiness in regard to public policy for this book. Giovannini is the minister of labor and social policies in the Italian government under Prime Minister Enrico Letta and played a formative role in steering the OECD to focus on well-being and progress in his role as chief statistician for the organization. He launched the Global Project on the Measurement of Progress in Societies, which fostered the setting up of numerous worldwide initiatives focused on the Beyond GDP movement. For his work on the measurement of societal well-being in 2010, he was awarded the Gold Medal of the President of the Republic of Italy, and has also been a member of the Commission on the Measurement of Economic Performance and Social Progress and chair of the Global Council on the Evaluation of Societal Progress established by the World Economic Forum. I asked Minister Giovannini how he first became involved in studying issues around well-being and happiness.

> In 2001, the OECD was running a project on measuring sustainable development. As an economist, I found this to be a fascinating effort. I was intrigued by the idea of integrating economics, social and environmental measures that also had an intergenerational component, and a long-term story.[21]

I asked the minister how he dealt with skeptics who may have thought studying well-being or happiness was impractical in light of the more financially focused metrics of the GDP.

> At the beginning it was very difficult to take these ideas forward, especially with the economists and statisticians. But this has changed for several reasons. Firstly, several governments are taking these types of measures very seriously, including the French, German, Australian, Japanese,

Korean, and Chinese leaders. Everybody understands that just increasing income is not enough. You have to look at all dimensions of life that include social and environmental factors.[22]

However, recessions and other economic issues do impact the study of well-being. As Minister Giovannini noted:

The idea of measuring and implementing happiness metrics can be very difficult to apply. During recessions a lot of people lose their jobs, which means their happiness decreases. So finding the right balance of policies, where you can look at all dimensions of a person's life, is essential in both emerging and developed countries. However, these types of crises also push citizens to ask for policies with greater justice to allow for fair distribution of resources, along with resources that are more sustainable. So now the measurement battle around well-being and happiness is almost won, but our next step is to create policies that can include these different elements.[23]

Mass Happiness

"Governments aren't put into place just to manage the GDP. Governments should make people better off." Daniel Hadley is the director of Somerstat, a program focused on analyzing municipal needs for the city of Somerville, Massachusetts, and providing forums for direct citizen participation in civic engagement. Beyond its fame as the home of Marshmallow Fluff, Somerville has become a leader in the usage of happiness indicator metrics to drive policy change. In the *New York Times* article "How Happy Are You? A Census Wants to Know," author John Tierney documents how residents were sent surveys asking people to rate both pragmatic aspects of their communities (schools, housing) as well as the beauty of the physical landscape. Overall, the city was trying to gauge the answer to the question, "Taking everything into account, how satisfied are you with Somerville as a place to live?"[24]

Daniel Gilbert, renowned Harvard University professor, social psychologist, and author of *Stumbling on Happiness*, helped Daniel and the staff at Somerville create the survey questions, which were also inspired by the work Prime Minister David Cameron has been doing in the United Kingdom with his Happiness Index. Daniel also utilized the groundbreaking work of the Knight Foundation and their Soul of the Community project to build Somerville's survey as he told me in an interview for *Hacking H(app)iness*:

> I borrowed from the best and specifically looked for questions that correlated with resident satisfaction. Nobody to my knowledge has combined a municipal survey with a happiness survey. We hoped we could mine the data and find out what municipal services could make people happy.[25]

Now that the city has collected two years' worth of data, Daniel and his team will start to be able to analyze trends in hopes of creating a Happiness Index that could be sharable with other cities such as Santa Monica that are also working to create measures of well-being to help citizens. While a snapshot of data from one year's survey is helpful, information from two surveys means the mayor's office can try to implement relevant policy change based on citizen input. And this idea is already working. In one simple yet charming example, Daniel had data that the number of trees in someone's neighborhood can affect people's happiness. So the city planted more trees and raised resident happiness as measured by survey response.

While policy change can get caught up in bureaucratic red tape or bipartisan rhetoric, it doesn't have to. The transparency from the Somerville surveys means the mayor's office will need to be responsive to citizens' requests in order to maintain trust and participation. But Daniel feels the results have been positive so far, and sees much more work to be done. From our interview:

> This framework is still in its infancy. I get excited about the future. I see every city doing some version of the Happiness Index. By 2030 we'll have metrics that will let us know that happiness shot way up in certain regions of the country. An average citizen can look at a map and see where happiness is the highest. Citizens in the future will be informed about where happiness is at its peak and why.[26]

As citizens we can take comfort in the fact that cities like Somerville are working to incorporate data that genuinely impacts our lives. Metrics that go beyond GDP don't just work because they offer theoretical promise. They also have to work when put into pragmatic practice.

So it's decided. While the GDP may have been a useful metric for a time, its fiscal-only focus ignores a number of issues central to accurate measurement and policy creation. It largely ignores women or people who stay at home with their kids but don't "produce value." While it is helpful to have any standard that the entire world agrees upon, it doesn't make sense to cling to a metric that was developed almost one hundred years ago in a completely different time.

So, gross domestic product? Thanks for playing. But now? Ba-BYE.

19

GETTING H(APP)Y

An excessive focus on happiness would seem to be almost disrespectful to the wide range of possible human emotions that lift us up, teach us, and make life rich and varied. A more thoughtful goal, or intention, or reason to try tracking mood, is simply to increase awareness. The act of pausing to check in with yourself about how you're feeling in different situations, as well as looking back to similar situations in the past, can help you see trends and influences on your mood that you may not ever have noticed.[1]

ROBIN BAROOAH AND ALEX CARMICHAEL

TO MOVE BEYOND GDP on a personal level, let's give it a new name: *gradual daily progress.*

Hacking H(app)iness is not supposed to be an easy fix. It's a process that begins with a bold declaration to radically examine and optimize the way you think about money, self-worth, and joy in your life. In my case, seeing my Klout score and realizing how others could broadcast data about my life spurred the journey that led to the writing of this book and the founding of the H(app)athon Project. The process hasn't been easy, but that's one of the main reasons it's been so utterly satisfying.

I hope you'll get to experience an epiphany in your life as I did. A clarifying moment where you're inspired to make a change and have a sense of direction on how to proceed is a blessing. My epiphany, however, came not too long after my father passed away.

While I wasn't looking for a radical life change, I had been in a state of deep introspection for a number of months dealing with my dad's death. I was open to receiving the epiphany when it came.

So to be clear: Hacking H(app)iness is not about "finding your happy place" or always being in a positive mood. It's about giving yourself permission to evaluate what brings you meaning and purpose. You *want* this process to be hard. You *want* it to get ugly, at least in terms of honoring a process that is *real*.

There is honor in seeking truth. I don't know what yours is. My goal in this chapter is to encourage you by providing some closing examples and stories to help you start exploring.

The Value of Values

Konstantin Augemberg is a statistician with a passion for quantifying his own life. His Measured Me blog and work is an "ongoing personal experiment in self-quantification and self-optimization" with an ultimate goal to "empirically demonstrate that any aspect of my everyday life can be quantified and logged on a regular basis, and that the knowledge from these numbers can be used to help me live better."[2] I interviewed him about his recent Hacking Happiness experiment,[3] which Konstantin was kind enough to say was partially inspired by the H(app)athon Project. It focused on analyzing which aspects of his life made him happy and why.

Do you genuinely think people can track their emotions or happiness?
First, it is important to understand the differences between measurement and tracking. Measurement is a process by which a certain construct (latent or tangent) is expressed in terms of numbers or categories. Tracking is a consistent, repetitive measurement of the construct in everyday settings, often "on the go." You can measure calories burned in the lab setting, in calorimetry labs, in a

hermetically sealed room. But if you want to track your calorie expenditure on a regular basis, every day, then tracking devices like BodyMedia would be your best choice.

Likewise, emotions, happiness, and other latent constructs can be measured objectively and numerically. I am not a specialist, but I would say you can detect happiness and emotional states by observing activity of different parts of your brain via a CAT or MRI scanner. However, devices that could enable you to track happiness or emotional states "on the go," in everyday life settings and relatively continuously, do not exist yet, at least to my knowledge. But you can still measure and track your happiness daily using short self-questionnaires. Even asking simple "How happy am I?" questions once or twice a day can lead to amazing discoveries, provided that you keep track of your answers.

What were the results of your Hacking H(app)iness experiment you were most surprised by? Encouraged by? And can other people replicate what you did and hack happiness?
The most surprising finding was how much living according to my personal values affects my happiness. In addition to recording how happy I am, I was recording how important some life priorities (family, money, career, friends, justice in the world, spiritual balance) were to me at a given point in time and then how satisfied I was with my attempt to live according to these values.

For instance, I would wake up in the morning and ask myself how happy I was. Then I would ask how important it was for me to earn a lot of money, have a successful career, have good relationships with family and my partner. Then I would ask myself how satisfied I was with my current financial situation, my career, and my relationships. Then I would repeat the process in the afternoon and evening. The experiment lasted one month.

Then I looked at the difference between expectations and reality for each of these life priorities and how these gaps were related to my happiness. I thought life priorities like money and career would have a considerable influence on my happiness. As it turns out, they had no impact whatsoever. But being able to express myself, being healthy, and being independent and spiritual were important predictors of my happiness. In other words, every time I felt like it was important for me to be creative and independent but was not able to express myself or act freely, my happiness level would decrease.

Other people can certainly replicate this experiment. I am not sure, however, that they will get similar results. Unlike in regular science, results of self-tracking experiments are not necessarily generalizable; what worked for me won't necessarily work for you. And that is all right, because that is the main goal of self-tracking and self-quantification: Analyze your own life to find your unique solutions to your own problems. And yes, if a person feels that he is unhappy, then he or she should definitely give "hacking" a try.[4]

I find Konstantin's experiment a fascinating example of focusing on currencies that have nothing to do with wealth. The fact that his happiness level decreased when he wasn't able to express his values is also compelling. While we all have to do things we don't want to in our lives, tracking our activities and noting their effects can help us prioritize how we want to spend our time.

The Billion People Project

Measuring your own life is a powerful motivator for happiness. Tracking your actions in aggregate with like-minded individuals can also greatly accelerate positive well-being.

The Billion People Project (BPP) is providing this type of opportunity. I interviewed Della and Carrie van Heyst of the Van Heyst Group in Boulder, Colorado, founders of this project, which is aimed at getting people to engage in planet-conscious actions that can minimize and reverse negative effects to the environment. As people get involved and take action, they are measured and broadcast in real time on the project's website and app. Tracking aggregate action becomes the inspiration for large-scale positive change. "Our goal is to bring together massive amounts of people to help the environment and move the dime on policy," notes Della.[5]

The Van Heyst Group is known for the high caliber of events they've hosted for more than forty years for clients like Cisco, *Fortune* magazine, and Equinix. Now they're leveraging their skills at creating passionate communities to increase people's happiness while reversing environmental erosion. Inspired to create an "action tank" versus a "think tank," the mother/daughter pair sense a pivotal shifting point regarding technology and how it can impact genuine human relationships. As Carrie pointed out in our interview:

> Tech has taken over too much of our lives. Teens are sitting next to each other and texting versus talking. When we first got exposed to the Internet, all we wanted to know was how we could get more connected to it. Now we're asking ourselves how we can get more connected to each other again. Can we move toward a happiness- or values-based economy? We need to do a check-in with ourselves and ask: What are our values and how can we express them?[6]

Having run over four hundred events around the world, the Van Heysts will be able to leverage key relationships with their friends in the tech and business communities to make the Billion People Project a reality. The project differs from other environmental campaigns in regard to its focus on data collection and participant's personal environmental impact on water, carbon, waste, air, and natural habitats—all leading to sustaining the health and happiness of the individual and the planet.

In the same way that companies are required to have offsets for any potential harm they cause to the environment, with the BPP, individuals can experience the tangible ways their actions hurt or help the earth. Where it may seem impossible to make global change as an individual, the Billion People Project will poignantly show aggregate impact. "My thought was, rather than just sit around, let's take action into our own hands," says Della. She continues:

> We, the people, can do this. Technology lets us scale our individual actions. And it's simple stuff—eliminate plastic in your life, change out an old heater. Walk more often than you drive. A lot of people are doing this, but they don't see the impact of what they're doing. We'll aggregate this in an effort to show how we're all connected as human beings.[7]

Upworthy and the Third Metric

I mentioned a while back how my father would ask all of his patients if they watched the eleven o'clock news. If they said yes, he would recommend that they stop watching. His point was not to try to keep people from facing reality but to help them shift their focus away from media that present news or information with certain biases. While it would take too long to discuss the nature of objectivity in journalism, it goes without saying, especially in the United States, that the top news stories on most shows focus on negative events. If you watch three local news stations in any market, for instance, you can even see the formula for most shows—two or three top stories typically focusing on generally negative events, followed by a "color piece" near the end of the broadcast highlighting a positive local event—a charity event, a remarkable pet, etc. Most people don't realize that even if the first few stories are presented in an objective light, the way the pieces are ordered is purposeful, designed to attract and keep viewers watching. While the formula is not necessarily diabolical, it's important to note how it has affected our overall consciousness, and also why late-night talk shows come immediately after the eleven o'clock news—we need something to laugh at quickly because we're so distressed by what we've just seen.

Upworthy (http://www.upworthy.com) was cofounded by Eli Pariser, whom I interviewed about his book *The Filter Bubble*. He and his team have done an amazing job of providing a refreshingly real and admittedly biased (toward the positive) framework for sharing stories intended to entertain, empower, and edify. Here's a bit of language from their "about" page:

> We're a mission-driven media company. We're not a newspaper—we'd rather speak truth than appear unbiased . . . But we do have a point of view. We're pro gay marriage, and we're anti child poverty. We think the media is horrible to women, we think climate change is real, and we think the government has a lot to learn from the Internet about efficiency, disruption, and effectiveness.[8]

I'm a firm believer that it's actually easier to be objective with reporting if you admit your biases upfront to your audience. I also believe in basic journalistic standards, such as giving two sides of a story, accurately citing sources, and so on. But it's the easiest thing in the world to veil your true opinion behind research you feature to prove your point. That's why I'm boldly telling you with this book that you're lying to yourself if you think the majority of modern news isn't weighing you down. While you can't control what happens in the world, or how it's reported, you are allowed to decide how and when you want to ingest it. And there's a difference between avoiding truth and being purposeful about which voices you bring into your life on a daily basis.

Here's one quick example of why I love the Upworthy site so much—a video by Rebecca Eisenberg[9] in response to some "old school, YouTube fat hate" she'd been receiving about her weight. In a little under three minutes she beautifully describes the difference between being fat and all the stigma attached to a person's size. She's smart, specific, brave, and bold, and offers an utterly refreshing take on weight issues versus the typical polarized "don't bully" versus "hater" debates we've heard for years. Beyond the fact that I've battled with being heavy for years and thereby sympathized with her views, since having the epiphany that launched this book and the H(app)athon Project, I crave and seek raw truth. It's so much more meaningful and satisfying than overt bias veiled in objectivity or a rampant worldview that has chosen to see the world through a negative lens.

The Third Metric is part of the *Huffington Post* and you can see it here: http://www.huffingtonpost.com/news/third-metric. The articles featured on this portion of the site are "redefining success beyond money and power" and reflect an overtly Beyond GDP mind-set meant to change the status quo surrounding ideas of how we work, live, and find well-being. As Arianna Huffington pointed out about the site (and the conferences focused on the same issues) in a recent *Chicago Tribune* article, "The motivation for these events is that it has become increasingly clear that the current model, in which success is equated with overwork, burnout, sleep deprivation and never seeing your family, isn't working. It's not working for women. It's not working for men. It's not working for companies, for any societies in which it's dominant or for the planet."[10]

There are a number of things I love about this site/conference. First, it honors women. I'm married to a woman and have a daughter and can speak from deep personal experience—women are awesome. It is beyond pitiful and ludicrous that in 2013 there should even be a need for a site/conference dedicated to women but sadly it's more needed than ever. However, what I appreciate about the site is its how-to focus regarding proactive ways to lower your stress or simply identify the paradigm of incessant productivity most of us feel equates to being successful. The *Huffington Post* also features an app/site called GPS for the Soul (http://www.huffingtonpost.com/gps-for-the-soul/) that provides similar proactive ways to measure and combat stress. The app even lets you measure your pulse by placing your finger on the lens of your mobile phone camera. After you get your pulse, you can see the videos or articles available on the site or create your own slide show from personal pictures to actively calm yourself down.

The Paradigm of Being Proactive

These are just a few samples of sites and voices designed to help you reorient your daily perspective on how positivity can actually be crafted in your life. You're allowed to reflect on what truly brings you meaning, and also understand how deeply your worth doesn't have to be focused on your wealth or outward image and influence.

This whole section of *Hacking H(app)iness* is about being proactive—promoting personal and public well-being versus just getting money and accumulating influence as a primary objective for your life. If, to quote Avner Offer, "the currency of well-being is attention," we all have to get better at spending time looking more deeply at ourselves while also regarding others and their needs as important as our own.

20

HACKING H(APP)INESS

I have learned to be comfortable with mystery.

DAVID W. HAVENS, M.D.

I DIDN'T REALIZE how much of this book was about my dad until I finished it.

He inspired it, as I pointed out in the introduction. And I've mentioned him a few times throughout the book. But for all the geektastic technology and economics I've learned about in my work on *Hacking H(app)iness*, I keep coming back to my dad.

- He did work he was built for.
- He got paid, but money was never the focus of his work.
- His life's calling was to listen and help others.

Put these three things together and you see a man who got paid to do what he loved. But did the fact he got paid alter the *value* he gave his patients? The value went beyond mere transaction. We were never rich, but I know Dad created a legacy of wealth for the people he cared for and their families.

After I got my driver's license in high school, I would often go to pick up my dad after work. Sitting in his worn leather armchair still redolent with the embedded scent of Borkum Riff tobacco from his pipe-smoking days, I could feel a palpable sense of deep emotion permeating the room. The experience was strikingly similar to the feeling I'd get for years as a professional actor in a theater after an audience had seen a play. Theater is therapy, as much for the actors as the audience. Scripts, lighting, costumes— they're all just pieces of a mirror we agree to look at together for a moment in time, giving us permission to reflect.

I received a letter from one of my dad's patients not long after he died. She told me how much he had meant to her in a deeply troubling time of her life. She gave me a glimpse of his life I had never seen. It was a precious gift, and it stands as a written testament for the shared experiences of thousands of other people my dad touched with his work. And his life.

We don't need to write everything down. Sometimes it can be exhausting or counterproductive to measure just for measuring's sake. But technology and science are helping to create ways to peer more deeply into our lives so we can let in some light to areas where we can inspire healing and growth. And people around the world are recognizing our inherent value has been supplanted for too long by the reckless pursuit of money, and that we're worth more than wealth.

The Heart of Hacking H(app)iness

As a review, here's what I've done my best to prove in this book:

- Data is getting personal.
- Happiness can be quantified and increased.
- The happiness economy is redefining wealth.

Here are the benefits I explained you'd gain from reading the book:

- Informed Choice
- Joyful Discovery
- The Currency of Connection

Here's how I've encouraged you to act in the context of mobile/modern technology, positive psychology, and evolved economic models focusing on shared value and balanced well-being:

- A—be Accountable
- P—be a Provider
- P—be Proactive

I've done my best to show you the following:

- Your personal data is more connected to the world than ever before.
- By measuring your life, you can optimize it and increase your happiness/well-being.
- By connecting your skills to actions that help others, the whole world gets h(app)y.

In short, I've tried to prove why your personal data counts and how tracking it can change the world. Now you know that I mean this literally and figuratively, and you can start the process now.

The Mystery

My dad's quote about learning to be comfortable with mystery wasn't a cop-out. He said it to me years ago, after we'd had a long argument about religion.

There was a time I planned on going to seminary to be a minister and I got caught up in analyzing Scripture scientifically in an effort to prove it was "true." When you're young in your faith (whatever the worldview), it's easy to think you can convince others to accept your beliefs if you have strong enough words. My dad, however, emphasized the importance of works and that people should know you from the fruits of your labor. He felt character was built, forged like the tools his blacksmith grandfather created when he was a boy.

I'm good with mystery. It leads to wonder. And awe. And humility.

So here's what I know, at the end of our journey together: Socrates said the unexamined life is not worth living.

Go see for yourself.

ACKNOWLEDGMENTS

I am blessed to know a trove of talented, brilliant, and gracious people who have lent their support in my writing of this book. For the others I've forgotten who should be in this list, my sincere apologies, and the pint is on me the next time we get together so I can thank you in person.

Thanks to Lynne D. Johnson for introducing me to my tireless literary agent, Carole Jelen, and her colleague Zachary Romano. Big shout-out to Andrew Yackira, my editor at Tarcher (Penguin Random House), for his constant encouragement and humor during the writing process. Thanks also to Gina Rizzo at Tarcher for her help in promoting the book, along with Nettie Reynolds, who helped me spread the word about *Hacking H(app)iness*. Rich Silivanch of Mutant Media is an amazing friend to me, this book, and the H(app)athon Project, and his compelling, attractive, and savvy branding and media creation (which includes the awesome logo and cover for this book) have provided fundamental help to my work. Kudos also to Wedge Martin of GeoPapyrus for his support.

Huge geeky thanks and shout-outs to Matt Silverman of Mashable, who provided me with the opportunity to create the series of work that inspired the articles that led to the writing of this book. Thanks also to Matt Petronzio, Stephanie Buck, and the rest of the team at Mashable who supported my articles and work. I also deeply appreciate the thoughts and wisdom of all of the interviewees I had the pleasure of learning from for this book, whether I spoke to them for Mashable or directly for *Hacking H(app)iness*. Thank you.

A great deal of research for this book came in unison with my work as founder of the H(app)athon Project. On that side of things, special thanks go to Adam Laughlin, Kat Houghton, and Sean Bohan, the team driving H(app)athon from the beginning. Their wisdom, expertise, and support are the reasons the Project has come to fruition. The Project also would not exist had it not been for the tireless efforts of Alexis Adair, with her gifts in research, positive psychology, and all things organizational help-ing me and H(app)athon make sense to the outside world. Other H(app)-athon peeps supporting me/the book include my advisers: William Hoffman of the World Economic Forum, Jon Hall of the United Nations Development Programme, Peter Johnson of the Van Heyst Group, Stephen J. Derezinski III of Infinium, and Peter Vander Auwera. These are the people providing essential outside perspective on our work to keep our vision on track.

Special thanks also go to Kat's team at ilumivu for the technology that launched the first version of our survey, and the team who created Open Mustard Seed, the heart of H(app)athon's transparent tech, including the amazing John Clippinger, Patrick Deegan, and the entire team at ID3. Piv-otal in helping us get the word out about our efforts is the gloriously self-less and gifted team at Pulp PR, chaired by Jessica Hasson, with hugely effective support from her colleague Mishri Bhatia. Also instrumental in sharing our work is our indefatigable social media team, led by Irene Koehler, the always-optimistic Debbie Miller, and Deanna Pitta. In Somer-ville, extreme thanks go to Daniel Hadley and his team at the mayor's office. Heartfelt thanks also to my friend and positive-future brother Chris Rezendes of INEX Advisors. Special thanks also to the amazingly warm Mario Chamorro, founder and CEO at Make It Happy and founder of The

Happy Post Project, and to all of his team. His support at our various H(app)athon Project events and introductions to people in the "happy space" have been invaluable.

I also want to warmly thank the other H(app)athonites around the world on our leadership committee and at other organizations that have added significantly to our work. These include Jason Williams, our global workshop coordinator and constantly supportive Twitter fanatic; Kristine Maudal, Even Fossen, Julie Kildahl, and the team at Brainwells in Oslo (Norway rules!); Juan Pablo Calderón in Bogotá and the team at HUB; Taichi Fujimoto from Happiness Architect and his team in Tokyo; Carlos Somohano and my old friend Stewart Townsend for their work in London; Stan Stalnaker of Ven for his numerous introductions; Manuel Kraus in Lisbon; Randall Krantz in Bhutan; Brian Barela, Brandon Cockrum, and the team in Indianapolis; Chris Heuer and his team in San Francisco; and the amazing Sunnie Southern in Cincinnati. Arthur Woods and his colleagues at Imperative deserve thanks for their help regarding how H(app)-athon can translate to the enterprise, and special thanks to KoAnn Skrzyniarz, Dimitar Vlahov, and Bart King of Sustainable Brands for all their support of H(app)athon and helping to host our San Diego event. Also a big shout-out to Kaliya (aka Identity Woman) for all of her amazing expertise regarding personal data and identity issues.

Profound thanks to Anil Sethi of Gliimpse for his financial support and entrepreneurial bonding during the H(app)athon process; Mario Chamorro of the Happy Post Project and his partnership in our workshop efforts; the gifted Peggy Kern and her colleagues at the University of Pennsylvania in the World Well-Being Project; Neal Lathia from the University of Cambridge for dealing with my numerous crazy and often late-night ideas; Tim Leberecht, fellow writer and artist type, who heard about H(app)athon soon after my Mashable article and has been so supportive ever since; Joshua Middleman, an early brainstormer and supporter in the H(app)athon process; Laura Musikanski from the Happiness Initiative for providing my earliest knowledge of all things related to happiness economics; J. P. Rangaswami of Salesforce.com for his ongoing wisdom and support of our work; plus Konstantin Augemberg of Measured Me; Dara Barlin of A Big Project; Mary Czerwinski of Microsoft; Margie Morris of

Intel; Susannah Fox of Pew Internet; Scott L. David; Johannes Eichstaedt; Lakshmi Arthi Krishnaswami; Kalev Leetaru; Amber Melhouse; Gregory Park, Ernesto Ramirez; Andrew Schwartz; Eiji Han Shimizu, Jason Sroka; Patrick Van Kessel; Kim Whittemore; Barbara Van Dahlen; Rick Cohen; Mark Anielski; Eimear Farrell; Shwen Gwee; Joel Johnson; and my dear friend David Richeson, who inspired me to try to make the H(app)athon Project a reality from the beginning. Thanks, Dave.

Thanks to my brother, Andy, for inspiring me to be a writer. And thanks to my mom. While my dad is mentioned a lot in this book, my mom is the one who has been my quasi-business manager, dear friend, and spiritual mentor for the better part of my entire life. I love you, Mom.

I always say that I married up regarding my wife, Stacy, and nothing could be truer. She is my best friend and was utterly supportive during the crazy packed months of parenting I was often absent for during the writing of this book.

Thanks to my son, Nate, and daughter, Sophie Joan, who make my life worth living. They're the primary reasons I want to reflect on my life more often, to spend time with them and Stacy versus staring at my iPhone.

And although the book is dedicated to him, he's worth mentioning again: My deepest and humblest thanks to a dad who was an artist with his silences as much as his words. Look forward to seeing you again, Pop. Until then, I pray my legacy can honor your work, which touched and helped so many lives.

NOTES

Introduction

1. David Bollier, *Power-Curve Society: The Future of Innovation, Opportunity and Social Equity in the Emerging Networked Economy,* Aspen Institute Communications and Society Program, 2013, http://www.aspeninstitute .org/sites/default/files/content/upload/Power-Curve-Society.pdf.
2. Steffan Keuer and Pernille Tranberg, *Fake It! Your Guide to Digital Self-Defense* (Amazon Digital Services, 2013).
3. Shane Green, CEO of Personal.com, interview with author, March 18, 2013.
4. *Protecting Consumer Privacy in an Era of Rapid Change: Recommendations for Businesses and Policymakers,* Federal Trade Commission Report, March 2012, http://ftc.gov/os/2012/03/120326privacyreport.pdf.
5. Federal Trade Commission, "FTC to Study Data Broker Industry's Collection and Use of Consumer Data," press release, December 18, 2012, http://www.ftc.gov/opa/2012/12/databrokers.shtm.
6. Sarah Perez, "Facebook Graph Search Didn't Break Your Privacy Settings, It Only Feels Like That," *TechCrunch,* February 4, 2013, http://techcrunch .com/2013/02/04/facebook-graph-search-didnt-break-your-privacy-settings-it-only-feels-like-that/.
7. Kaliya (aka Identity Woman), executive director, Personal Data Ecosystem Consortium, interview with author, August 2, 2013.
8. David Streitfeld, "Google Concedes That Drive-By Prying Violated Privacy," *New York Times,* March 12, 2013, http://www.nytimes.com/2013/ 03/13/technology/google-pays-fine-over-street-view-privacy-breach.html? pagewanted=all.

9. *Privacy and Security,* Edelman's Data Security and Privacy Group, 2012, http://datasecurity.edelman.com/wp-content/uploads/2012/03/Data-Security-Privacy-Executive-Summary.pdf.
10. John Helliwell, Richard Layard, and Jeffrey Sachs, eds., *World Happiness Report,* Earth Institute, April 2, 2012, http://www.earth.columbia.edu/articles/view/2960.
11. Michelle Nicole Burns, Mark Begale, Jennifer Duffecy, Darren Gergle, Chris J Karr, Emily Giangrande, and David C Mohr, "Harnessing Context Sensing to Develop a Mobile Intervention for Depression," *Journal of Medical Internet Research,* August 12, 2011, http://www.ncbi.nlm.nih.gov/pmc/articles/PMC3222181/.
12. http://emotionsense.org.
13. Kiran K. Rachuri, Mirco Musolesi, Cecilia Mascolo, Peter J. Rentfrow, Chris Longworth, and Andrius Aucinas, "Emotion Sense: A Mobile Phones–based Adaptive Platform for Experimental Social Psychology Research," University of Cambridge Computer Laboratory, 2013, http://www.cl.cam.ac.uk/~cm542/papers/Ubicomp10.pdf.
14. John C. Havens, "The Impending Social Consequences of Augmented Reality," *Mashable,* February 8, 2013, http://mashable.com/2013/02/08/augmented-reality-future/.
15. Shirley S. Wang, "Is Happiness Overrated?" *The Wall Street Journal,* March 15, 2011, http://online.wsj.com/article/SB10001424052748704893604576200471545379388.

Chapter One

1. Coert Visser, "New Light on the Origin of the Scaling Question—'The Cantril Self-Anchoring Striving Scale,'" *The Doing What Works Blog,* May 22, 2010, http://solutionfocusedchange.blogspot.com/2010/05/new-light-on-origin-of-scaling-question.html.
2. http://www.slideshare.net/swipp/rateocracy-when-everyone-and-everything-is-rated.
3. Christopher Carfi, VP of platform products for Swipp, interview with author, June 3, 2013.
4. Stephanie Pappas, "Facebook with Care: Social Networking Site Can Hurt Self-Esteem," *LiveScience,* February 6, 2012, http://www.livescience.com/18324-facebook-depression-social-comparison.html.
5. Howard Gardner, Hobbs Professor of Cognition and Education, Harvard Graduate School of Education, interview with author, June 9, 2013.
6. https://www.youtube.com/watch?v=hL4lSavSepc.
7. Cassie Shortsleeve, "De-Stress in 3 Seconds," *Men's Health,* August 8, 2012, http://news.menshealth.com/de-stress-in-3-seconds/2012/08/08/.
8. David Talbot, "Wrist Sensor Tells You How Stressed Out You Are," *MIT Technology Review,* December 20, 2012, http://www.technologyreview.com/news/508716/wrist-sensor-tells-you-how-stressed-out-you-are/.
9. Bianca Bosker, "Affectiva's Emotion Recognition Tech: When Machines Know What You're Feeling," *Huffington Post,* December 24, 2012, http://www.huffingtonpost.com/2012/12/24/affectiva-emotion-recognition-technology_n_2360136.html.
10. http://mashable.com/category/social-tv/.

Chapter Two

1. Rachel Botsman, "The Currency of the New Economy Is Trust," TED Talk video, June 2012, http://www.ted.com/talks/rachel_botsman_the_currency_of_the_new_economy_is_trust.html.
2. Malgorzata Wozniacka and Snigdha Sen, "Credit Scores: What You Should Know About Your Own," *Frontline,* November 23, 2004, http://www.pbs.org/wgbh/pages/frontline/shows/credit/more/scores.html.

3. John C. Havens, "Why Social Accountability Will Be the New Currency of the Web," *Mashable*, July 28, 2011, http://mashable.com/2011/07/28/social-media-influence-accountability/.

4. Ibid.

5. Sandy Pentland, *Sensible Organization*, MIT Sloan School of Management, 2006, http://hd.media.mit.edu/tech-reports/TR-602.pdf.

6. Sandy Pentland, "Sensible Organizations: Technology and Methodology for Automatically Measuring Organizational Behavior," *IEEE Transactions on Systems, Man, and Cybernetics: Part B: Cybernetics* 39, no. 1 (February 2009): 43–55, http://realitycommons.media.mit.edu/Sensible_Organizations.pdf.

7. Klint Finley, "What if Your Boss Tracked Your Sleep, Diet, and Exercise?" *Wired*, April 17, 2013, http://www.wired.com/wiredenterprise/2013/04/quantified-work-citizen/.

8. Quinn Simpson, user experience director, Citizen, interview with author, April 23, 2013.

9. Jon Bowermaster, "Thanks to Plastic, Your Chances of Finding Nemo Just Got Way Worse," TakePart.com, April 19, 2013, http://www.takepart.com/article/2013/04/19/plastic-pollution-in-the-ocean.

10. "The Rise of the Sharing Economy," *Economist*, March 9, 2013, http://www.economist.com/news/leaders/21573104-internet-everything-hire-rise-sharing-economy.

Chapter Three

1. Senator Al Franken, interview with author, April 26, 2013.

2. "Under 13 Year Olds on Facebook: Why Do 5 Million Kids Log in if Facebook Doesn't Want Them To?" *Huffington Post*, September 19, 2012, http://www.huffingtonpost.com/2012/09/19/under-13-year-olds-on-facebook_n_1898560.html.

3. William Hoffman, director of the World Economic Forum's Information, interview with author, March 23, 2013.

4. *Unlocking the Value of Personal Data: From Collection to Usage*, Boston Consulting Group, in conjunction with the World Economic Forum, February 2013, http://www3.weforum.org/docs/WEF_IT_UnlockingValuePersonalData_CollectionUsage_Report_2013.pdf.

5. Latanya Sweeney, "Simple Demographics Often Identify People Uniquely," Data Privacy Working Paper 3, Carnegie Mellon University, 2000, http://dataprivacylab.org/projects/identifiability/paper1.pdf.

6. *Unlocking the Value of Personal Data,* Boston Consulting Group.

7. Fatemeh Khatibloo, *Personal Identity Management: Preparing for a World of Consumer-Managed Data*, Forrester Research, September 30, 2011, http://www.forrester.com/Personal+Identity+Management/fulltext/-/E-RES60322?docid=60322.

8. Shane Green, CEO of Personal.com, interview with author, March 8, 2013.

9. Ibid.

10. Michael Fertik, CEO of Reputation.com, interview with author, April 9, 2013.

11. David Bollier, *Power-Curve Society: The Future of Innovation, Opportunity and Social Equity in the Emerging Networked Economy*, Aspen Institute Communications and Society Program, 2013, http://www.aspeninstitute.org/sites/default/files/content/upload/Power-Curve-Society.pdf.

12. Senator Al Franken, interview with author, April 23, 2013.

Chapter Four

1. Tim O'Reilly, quoted in Mac Slocum, "What Lies Ahead: Data," *O'Reilly Radar*, December 27, 2010, http://radar.oreilly.com/2010/12/2011-data.html.

2. R. W. Picard, "Affective Computing," MIT Media Laboratory Perceptual Computing Section Technical Report 321, April 1995, https://www.pervasive.jku.at/Teaching/_2009SS/SeminarausPervasiveComputing/Begleitmaterial/Related%20Work%20%28Readings%29/1995_Affective%20computing_Picard.pdf.

3. Karen Weintraub, "But How Do You Really Feel? Someday the Computer May Know," *New York Times*, October 15, 2012, http://www.nytimes.com/2012/10/16/science/affective-programming-grows-in-effort-to-read-faces.html?pagewanted=all&gwh=637D4752C266669B2E2F0CA12CE24454.

4. Barbara Dunk and Kevin Doughty, "The Aztec Project: Providing Assistive Technology for People with Dementia and Their Carers in Croydon," paper presented at Laing & Buisson 2006 Telecare & Assistive Technology Conference, Cavendish Conference Centre, London, January 18, 2006, http://www.alzheimers-support.com/downloads/case-study-the-aztec-project.pdf.

5. Dimitris Mandiliotis, Kostas Toumpas, Katerina Kyprioti, and Kiki Kaza, *Symbiosis Alzheimer's Support: An Innovative Approach to Alzheimer's Support*, Aristotle University of Thessaloniki, http://www.symbiosis.ee.authgr/e-Symbiosis/Welcome_files/Symbiosis%20Additional%20Application%20Material%20IC%20Grants%202012.pdf.

6. Kat Houghton, cofounder of Ilumivu, interview with author, July 27, 2013.

7. Ibid.

8. Mary Czerwinski, research manager, VIBE Research Group, Microsoft, interview with author, August 5, 2013.

9. Iggy Fanlo, cofounder and CEO of Lively, interview with author, April 4, 2013.

10. Ibid.

11. Patrick Meier, "MatchApp: Next Generation Disaster Response App?" *iRevolution Blog*, February 27, 2013, http://irevolution.net/2013/02/27/matchapp-disaster-response-app/.

12. Patrick Meier, "Jointly: Peer-to-Peer Disaster Recovery App," *iRevolution Blog*, April 24, 2013, http://irevolution.net/2013/04/24/jointly-app/.

13. Patrick Meier, "MatchApp."

14. Margaret Morris, "The New Sharing of Emotions," TED Talk video, April 2013, http://www.youtube.com/watch?v=FWdTVmWhAEs.

15. Margaret E. Morris, Qusai Kathawala, Todd K. Leen, Ethan E. Gorenstein, Farzin Guilak, Michael Labhard, and William Deleeuw, "Mobile Therapy: Case Study Evaluations of a Cell Phone Application for Emotional Self-Awareness," *Journal of Medical Internet Research* 12, no. 2 (April–June 2010); e10, http://www.jmir.org/2010/2/e10/.

16. Margaret Morris, clinical psychologist and senior researcher at Intel, interview with author, August 26, 2013.

17. John C. Havens, "The Value of a Happiness Economy," *Mashable*, June 13, 2012, http://mashable.com/2012/06/13/happiness-economy/.

18. Bart King, "Run Up to SB '13: John Havens Crowdsourcing Happiness to Save the World," Sustainable Brands website, May 7, 2013, http://www.sustainablebrands.com/news_and_views/blog/run-sb13-john-havens-crowdsourcing-happiness-save-world.

Chapter Five

1. "Wearable Computing Devices, Like Apple's iWatch, Will Exceed 485 Million Annual Shipments by 2018," ABI Research website, February 21, 2013, http://www.abiresearch.com/press/wearable-computing-devices-like-apples-iwatch-will.

2. Gary Wolf, "The Data-Driven Life," *New York Times Magazine*, April 28, 2010, http://www.nytimes.com/2010/05/02/magazine/02self-measurement-t.html?pagewanted=all&gwh=BA49EBFD810CB1BA3B78ED44071DE9CB.

3. Antonio Regalado, "Stephen Wolfram on Personal Analytics," *MIT Technology Review*, May 8, 2013, http://www.technologyreview.com/news/

514356/stephen-wolfram-on-personal-analytics/?utm_campaign=
newsletters&utm_source=newsletter-daily-all&utm_medium=email&utm
_content=20130509.

4. Susannah Fox and Maeve Duggan, *Tracking for Health,* Pew Internet &
American Life Project, January 28, 2013, http://www.pewinternet.org/
Reports/2013/Tracking-for-Health.aspx.
5. https://twitter.com/RachelleDiGreg.
6. http://quantifiedself.com/guide/.
7. Ari Meisel, "Don't Try to Prioritize, Work on Your Timing," *The Blog of Ari
Meisel,* March 20, 2012, http://www.lessdoing.com/2012/03/20/
dont-try-to-prioritize-work-on-your-timing/.
8. John C. Havens, "How Big Data Can Make Us Happier and Healthier,"
Mashable, October 8, 2012, http://mashable.com/2012/10/08/
the-power-of-quantified-self/.

Chapter Six

1. Chris Rezendes, "IOT: Grandest Opportunity, Most Stubborn Challenges,"
Business Weekly, April 29, 2013, http://www.businessweekly.co.uk/blog/
business-weekly-guest-blog/15338-iot-grandest-opportunity-most-stubborn-
challenges.
2. Anne Hart, "How Trees Contribute to Health by Producing Oxygen and
Lowering Cortisol Levels," *Examiner,* August 6, 2012, http://www.examiner
.com/article/how-trees-contribute-to-health-by-producing-oxygen.
3. Sarah Zielinski, "Going to the Park May Make Your Life Better,"
Smithsonian, April 22, 2011, http://blogs.smithsonianmag.com/science/
2011/04/going-to-the-park-may-make-your-life-better/.
4. John C. Havens, "The Impending Social Consequences of Augmented
Reality," *Mashable,* February 8, 2013, http://mashable.com/2013/02/08/
augmented-reality-future/.
5. "Impaired Driving: Get the Facts," Centers for Disease Control and
Prevention website, April 17, 2013, http://www.cdc.gov/motorvehiclesafety/
impaired_driving/impaired-drv_factsheet.html.
6. M. Mobeen Khan, executive director of mobility marketing, AT&T
Business, interview with the author, May 2, 2013.
7. Bill Zujewski, CMO and EVP of product strategy for Axeda Corporation,
interview with author, April 15, 2013.
8. Liat Ben-Zur, senior director of product management, Qualcomm,
interview with author, April 4, 2013.
9. Gartner, Inc., "Gartner Identifies the Top Ten Strategic Technology Trends
for 2013," press release, October 23, 2012, http://www.gartner.com/
newsroom/id/2209615.

Chapter Seven

1. Greg Satell, "5 Trends That Will Drive the Future of Technology," *Forbes,*
March 12, 2013, http://www.forbes.com/sites/gregsatell/2013/03/12/
5-trends-that-will-drive-the-future-of-technology/.
2. "Algorithm," *Wikipedia,* accessed 06:29, July 23, 2013, http://en.wikipedia
.org/wiki/Algorithm.
3. John Tierney, "A Match Made in the Code," *New York Times,* February 11,
2013, http://www.nytimes.com/2013/02/12/science/skepticism-as-
eharmony-defends-its-matchmaking-algorithm.html?pagewanted=all&_r=
1&&gwh=943C82A4AE35C31D744677CCB7C27153.
4. Eli Pariser, *The Filter Bubble: How the New Personalized Web Is Changing
What We Read and How We Think* (New York: Penguin Press, 2011), 15.
5. Greg Linden, former principal engineer at Amazon, interview with author,
April 2, 2013.

6. Yue Wang, "More People Have Cell Phones Than Toilets, U.N. Study Shows," *Time*, March 25, 2013, http://newsfeed.time.com/2013/03/25/more-people-have-cell-phones-than-toilets-u-n-study-shows/#ixzz2ZrdXJp52.
7. *A Roadmap to U.S. Robotics: From Internet to Robotics* 2013 edition, the Robotics Virtual Organization, March 20, 2013, http://robotics-vo.us/sites/default/files/2013%20Robotics%20Roadmap-rs.pdf.
8. Alexander Reben, creator of Blabdroid, interview with author, April 3, 2013.
9. Güven Güzeldere and Stefano Franchi, "Dialogues with Colorful Personalities of Early AI," *Stanford Electronic Humanities Review* 4, no.2 (July 24, 1995), http://www.stanford.edu/group/SHR/4-2/text/dialogues.html.
10. Jeffrey R. Young, "Programmed for Love," *The Chronicle of Higher Education*, January 14, 2011, http://chronicle.com/article/Programmed-for-Love-The/125922/.
11. Filmmaker Brett Hoff, interview with author, April 3, 2013.
12. Angela Watercutter, "Neuroscientists Measure Brain Activity in *Love Competition*," *Wired*, February 14, 2012, http://www.wired.com/underwire/2012/02/love-competition/.
13. Alexander Peben interview.

Chapter Eight

1. Corien Prins, "When Personal Data, Behavior, and Virtual Identities Become a Commodity: Would a Property Rights Approach Matter?" *SCRIPTed* 3–4 (2006): 270, doi: 10.2966/scrip.030406.270, http://www2.law.ed.ac.uk/ahrc/script-ed/vol3-4/prins.asp.
2. Alexis C. Madrigal, "How Much Is Your Data Worth? Mmm, Somewhere Between Half a Cent and $1,200," *Atlantic*, March 19, 2012, http://www.theatlantic.com/technology/archive/2012/03/how-much-is-your-data-worth-mmm-somewhere-between-half-a-cent-and-1-200/254730/.
3. Corien Prins, "Personal Data."
4. John Gantz and David Reinsel, "The Digital Universe in 2020: Big Data, Bigger Digital Shadows, and Biggest Growth in the Far East," International Data Corporation, December 2012, http://idcdocserv.com/1414.
5. Rufus Pollock, "Forget Big Data, Small Data Is the Real Revolution," *The Guardian*, April 25, 2013, http://www.guardian.co.uk/news/datablog/2013/apr/25/forget-big-data-small-data-revolution.
6. Michael Schroeck, Rebecca Shockley, Dr. Janet Smart, Professor Dolores Romero-Morales, and Professor Peter Tufano, *Analytics: The Real-World Use of Big Data*, IBM Global Business Services, http://public.dhe.ibm.com/common/ssi/ecm/en/gbe03519usen/GBE03519USEN.PDF.
7. Claire Cain Miller, "Data Science: The Numbers of Our Lives," *New York Times*, April 11, 2013.
8. John C. Havens, "Big Data's Value Lies in Self-Regulation," *Mashable*, February 25, 2013, http://mashable.com/2013/02/25/big-data-self-regulation/.
9. Ibid.
10. Mark Bonchek, "Little Data Makes Big Data More Powerful," *Harvard Business Review Blog*, May 3, 2013, http://blogs.hbr.org/cs/2013/05/little_data_makes_big_data_mor.html.
11. Patrick Tucker, "Has Big Data Made Anonymity Impossible?" *MIT Technology Review*, May 7, 2013, http://www.technologyreview.com/news/514351/has-big-data-made-anonymity-impossible/.
12. Havens, "Big Data's Value Lies in Self-Regulation."
13. http://www.tictrac.com.

Chapter Nine

1. Dan Farber, "The Next Big Thing in Tech: Augmented Reality," CNET, June 7, 2013, http://news.cnet.com/8301-11386_3-57588128-76/the-next-big-thing-in-tech-augmented-reality/.

2. John C. Havens, "Self-Screening," *Media*, April 13, 2012, http://www
.mediapost.com/publications/article/172480/self-screening.html#axzz2
THigoMbB.
3. Brian X. Chen, "If You're Not Seeing Data, You're Not Seeing," *Wired*,
August 25, 2009, http://www.wired.com/gadgetlab/2009/08/
augmented-reality/.
4. John C. Havens, "Augmented Reality: What Marketers Need to Know,"
iMedia Connection, December 2, 2009, http://www.imediaconnection.com/
25256.asp.
5. Todd Wasserman, "How Google Glass Could Change Advertising,"
Mashable, January 23, 2013, http://mashable.com/2013/01/23/
google-glass-advertising/.
6. Kristina Grifantini, "GM Develops Augmented Reality Windshield," *MIT
Technology Review*, March 17, 2010, http://www.technologyreview.com/
view/418080/gm-develops-augmented-reality-windshield/.
7. https://www.youtube.com/watch?v=PKdFqUgm86k.
8. https://www.youtube.com/watch?v=zPfUN4ffssU.
9. https://vimeo.com/46304267.
10. https://vimeo.com/12899347.
11. Wedge Martin, cofounder of GeoPapyrus, interview with author,
May 14, 2013.
12. John C. Havens, "Who Owns the Advertising Space in an Augmented
Reality World?" *Mashable*, June 6, 2011, http://mashable.com/2011/06/06/
virtual-air-rights-augmented-reality/.
13. Chris Cameron, "Are We Entering the Age of Augmented Trademark
Infringement?" *ReadWrite*, July 6, 2010, http://readwrite.com/2010/07/06/
are_we_entering_the_age_of_augmented_trademark_infringement#awesm
=~ocqpE9IwkrwWdN.
14. Senator Al Franken, interview with author, April 26, 2013.
15. "Sen. Franken Calls for End to Disturbing Consumer Tracking Trend," Al
Franken website, April 1, 2013, http://www.franken.senate.gov/?p=press
_release&id=2341.
16. Senator Al Franken interview.
17. Brian Wassom, "FTC Issues Best Practices for Facial Recognition Privacy,"
Wassom.com Blog, October 31, 2012, http://www.wassom.com/ftc-issues
best-practices-for-facial-recognition-privacy.html.
18. Brian Wassom, partner, Honigman Miller Schwartz and Cohn LLP,
interview with author, June 2, 2013.
19. Robert Gordon, chief strategy officer, APX Labs, interview with author,
April 11, 2013.
20. https://www.youtube.com/watch?feature=player_embedded&v=
yRrdeFh5-io#!.

Chapter Ten
1. J. P. Rangaswami, quoted by Venessa Miemis in "The Bank of Facebook:
Currency, Identity, Reputation," *Forbes*, April 4, 2011, http://www.forbes
.com/sites/venessamiemis/2011/04/04/the-bank-of-facebook-currency
identity-reputation/.
2. Sarah Kessler, "What If We Thought More Often About Being Tracked
Online? Man Stalks Himself to Find Out," *Fast Company*, May 10, 2013,
http://www.fastcompany.com/3009602/what-if-we-thought-more-often
about-being-tracked-online-man-stalks-himself-to-find-out.
3. Eli Pariser, author of *The Filter Bubble*, interview with author, April 16, 2013.
4. John C. Havens, "Why Social Accountability Will Be the New Currency of
the Web," *Mashable*, July 28, 2011, http://mashable.com/2011/07/28/
social-media-influence-accountability/.
5. "Elinor Ostrom," *Economist*, June 30, 2012, http://www.economist.com/
node/21557717.

6. Kim-Mai Cutler, "Facebook's Payments Revenue Feels Some Heat, Declines 9% From Last Quarter," *TechCrunch*, October 23, 2012, http://techcrunch.com/2012/10/23/facebook-payments-revenue/.

7. Gartner, Inc., "Gartner Says Google and Facebook May Not Be the Banks of the Future, But They Will Help Shape the Future of the Financial Services Industries," press release, October 9, 2012, http://www.gartner.com/newsroom/id/2191115.

8. Samantha Murphy, "Only 10% of Americans Say They Would Wear Google Glass," *Mashable*, May 15, 2013, http://mashable.com/2013/05/15/google-glass-study/.

9. Hayley Tsukayama, "The Circuit: Amazon Introduces Its Own Currency, Amazon Coins," *Washington Post*, May 13, 2013, http://www.washingtonpost.com/blogs/post-tech/post/the-circuit-amazon-introduces its-own-currency-amazon-coins/2013/05/13/dd41a938-bbe1-11e2-97d4-a479289a31f9_blog.html.

10. "Amazon Coins v Trillion-Dollar Coins," *Economist*, February 7, 2013, http://www.economist.com/blogs/democracyinamerica/2013/02/value-and-virtual-world.

11. J. P. Rangawami of Salesforce.com, interview with author, January 21, 2013.

12. Paula Span, "A Volunteer Army of Caregivers," *New York Times*, March 28, 2013, http://newoldage.blogs.nytimes.com/2013/03/28/a-volunteer-army-of-caregivers/.

13. *Virtual Currency Schemes*, European Central Bank, October 2012, http://www.ecb.int/pub/pdf/other/virtualcurrencyschemes201210en.pdf.

14. Ibid.

15. https://www.youtube.com/watch?v=98RYtKQaWcA.

16. Phil Windley, "The Digital Asset Grid Session at SIBOS," *Technometria Blog*, November 1, 2012, http://www.windley.com/archives/2012/11/the digital_asset_grid_session_at_sibos.shtml.

17. John C. Havens, "Social Accountability."

18. Ibid.

Chapter Eleven

1. Michael E. Porter and Mark R. Kramer, "Creating Shared Value," *Harvard Business Review* (January–February 2011), http://www.hks.harvard.edu/m-rcbg/fellows/N_Lovegrove_Study_Group/Session_1/Michael_Porter _Creating_Shared_Value.pdf.

2. Ibid.

3. John C. Havens, "Social Responsibility: It's Not Just for Brands Anymore," *Mashable*, October 24, 2011, http://mashable.com/2011/10/24/social-responsibility-influence/.

4. Adam M. Grant, author of *Give and Take: A Revolutionary Approach to Success*, interview with author, May 14, 2013.

5. Ibid.

Chapter Twelve

1. Andrew Benett and Ann O'Reilly, *Consumed: Rethinking Business in the Era of Mindful Spending* (New York: Palgrave Macmillan, 2010), ix.

2. https://www.youtube.com/watch?v=SOnkv76rNL4.

3. Association of Psychological Science, "Consumerism and Its Antisocial Effects Can Be Turned On—or Off," press release, April 9, 2012, http://www.psychologicalscience.org/index.php/news/releases/consumerism-and-itsantisocial-effects-can-be-turned-onor-off.html.

4. Ibid.

5. John C. Havens, "The Impending Social Consequences of Augmented Reality," *Mashable*, February 8, 2013, http://mashable.com/2013/02/08/augmented-reality-future/.

6. "Savoring," *Positive Psychology News Daily Blog*, http://positivepsychologynews.com/search-by/image-maps/positive-emotions/savoring.

Chapter Thirteen

1. Adam Smith, *The Theory of Moral Sentiments* (London, 1759).
2. Avner Offer, "Between the Gift and the Market: The Economy of Regard," *Economic History Review* L, 3 (1997): 450–476, http://economics.ouls.ox.ac.uk/10489/1/gift3.pdf.
3. "Adam Smith," *Wikipedia*, accessed 08:24, July 26, 2013, http://en.wikipedia.org/wiki/Adam_Smith.
4. Avner Offer, *The Challenge of Affluence: Self-Control and Well-Being in the United States and Britain since 1950* (New York: Oxford University Press, 2006), 77.
5. Avner Offer, interview with author, March 15, 2013.
6. Barbara Fredrickson, "Your Phone vs. Your Heart," *New York Times*, March 23, 2013, http://www.nytimes.com/2013/03/24/opinion/sunday/your-phone-vs-your-heart.html.
7. Robin Dunbar, "How Many Friends Does One Person Need?" University of Oxford, February 4, 2010, http://www.ox.ac.uk/media/books/how_many_friends.html.
8. Kathleen Kennedy Townsend, "The Pursuit of Happiness: What the Founders Meant—And Didn't," *Atlantic*, June 20, 2011, http://www.theatlantic.com/business/archive/2011/06/the-pursuit-of-happiness-what-the-founders-meant-and-didnt/240708/.
9. Ibid.
10. Kathleen Kennedy Townsend, interview with author, June 17, 2013.

Chapter Fourteen

1. http://www.happify.com.
2. http://www.ippanetwork.org/faq/.
3. Martin Seligman, "The New Era of Positive Psychology," TED Talk video, February 2004, http://www.ted.com/talks/martin_seligman_on_the_state_of_psychology.html.
4. "Positive Psychology," *Wikipedia*, accessed 08:40, July 26, 2013, http://en.wikipedia.org/wiki/Positive_psychology.
5. Martin E. P. Seligman, "What Is Well-Being?" *Authentic Happiness Blog*, April 2011, http://www.authentichappiness.sas.upenn.edu/newsletter.aspx?id=1533.
6. http://youtu.be/vGc9pWan-FY?t=1m6s.
7. Martin Seligman, *Flourish: A Visionary New Understanding of Happiness and Well-being* (New York: Free Press, 2011), 20.
8. Jenn Lim, CEO, Delivering Happiness at Work, interview with author, May 13, 2013.
9. Barb Sanford, "The High Cost of Disengaged Employees," *Gallup Business Journal*, April 15, 2002, http://businessjournal.gallup.com/content/247/the-high-cost-of-disengaged-employees.aspx.
10. Marise Schot, head of the Happiness Lab at the Waag Society in Amsterdam, interview with author, April 29, 2013.
11. Jenn Lim, interview with author, May 13, 2013.
12. Kristine Maudal, partner at Brainwells in Oslo, Norway; interview with author, June 2, 2013.
13. Dacher Keltner, "The Compassionate Instinct," Greater Good website, spring 2004, http://greatergood.berkeley.edu/article/item/the_compassionate_instinct.
14. Ibid.

15. https://vimeo.com/51140422.
16. Eiji Han Shimizu, filmmaker, publisher, and workshop producer, interview with author, July 27, 2013.
17. John F. Helliwell and Haifang Huang, "Comparing the Happiness Effects of Real and On-line Friends," NBER Working Paper No. 18690, National Bureau of Economic Research, January 2013, http://nber.org/papers/w18690.

Chapter Fifteen
1. Viktor E. Frankl, *Man's Search for Meaning* (Boston: Beacon Press, 2006), xiv–xv.
2. https://www.youtube.com/watch?v=XBR8c6qvHMc.
3. Mihaly Csikszentmihalyi, *Flow: The Psychology of Optimal Experience* (New York: Harper & Row, 1990), 2.
4. Csikszentmihalyi, *Flow, 3*.
5. Mihaly Csikszentmihalyi, "Finding Flow," *Psychology Today*, July 1, 1997, http://www.psychologytoday.com/articles/199707/finding-flow.
6. https://vimeo.com/62071609.
7. Brenda Milner and Carol Tavris, "Inside the Psychologist's Studio," from the 24th Association for Psychological Science Convention, Chicago, filmed May 26, 2012, https://vimeo.com/50697025.
8. Steve Mann, Jason Nolan, and Barry Wellman, "Sousveillance: Inventing and Using Wearable Computing Devices for Data Collection in Surveillance Environments," *Surveillance & Society* 1, no. 3 (2003): 331–55, http://www.surveillance-and-society.org/articles1%283%29/sousveillance.pdf.
9. Oskar Kalmaru, cofounder of Memoto, interview with author, February 26, 2013.
10. Niclas Johansson, "Lifeloggers: Meet the Reporter, Amanda Alm," *Memoto Blog*, May 23, 2013, http://blog.memoto.com/2013/05/amanda-alm-the-reporter-of-lifeloggers/.
11. Jane Sarasohn-Kahn, *Making Sense of Sensors: How New Technologies Can Change Patient Care*, California HealthCare Association, February 2013, http://www.chcf.org/~/media/MEDIA%20LIBRARY%20Files/PDF/M/PDF%20MakingSenseSensors.pd.
12. Csikszentmihalyi, "Finding Flow."
13. Ibid.
14. Ben Waber, "Augmenting Social Reality in the Workplace," *MIT Technology Review*, May 15, 2013, http://www.technologyreview.com/news/514371/augmenting-social-reality-in-the-workplace/.
15. Lee Drutman, "Get Politically Engaged, Get Happy?" *Pacific Standard*, February 14, 2010, http://www.psmag.com/politics/get-politically-engaged-get-happy-8307/.
16. Malte Klar and Tim Kasser, "Some Benefits of Being an Activist: Measuring Activism and Its Role in Psychological Well-Being," *Political Psychology* 30, no. 5 (October 2009): 755–77, http://d.yimg.com/kq/groups/20066774/509269522/name/Activism.pdf.
17. Sara Armstrong, "The Key to Learning: A Place for Meaningful Academic Exploration," *Edutopia*, April 11, 2002, http://www.edutopia.org/key-to-learning-place-for-meaningful-academic-exploration.
18. "Mihaly Csikszentmihalyi: Motivating People to Learn," *Edutopia*, April 11, 2002, http://www.edutopia.org/mihaly-csikszentmihalyi-motivating-people-learn.

Chapter Sixteen
1. Steven Pinker, *The Blank Slate: The Modern Denial of Human Nature* (New York: Viking, 2002), 259.

2. Dacher Keltner, "The Compassionate Instinct," Greater Good website, Spring 2004, http://greatergood.berkeley.edu/article/item/the_compassionate_instinct.
3. Sonja Lyubomirsky, *The How of Happiness: A New Approach to Getting the Life You Want* (New York: Penguin Press, 2008), 22.
4. Lyuborsky, *The How of Happiness,* 128.
5. http://www.randomactsofkindness.org/kindness-stories/1506-gas-station.
6. John Helliwell, professor emeritus, Vancouver School of Economics, interview with author, April 19, 2013.
7. Dana Klisanin, psychologist, interview with author, June 12, 2013.
8. Rob Stein, "Happiness Can Spread Among People Like a Contagion, Study Indicates," *Washington Post,* December 5, 2008, http://www.washingtonpost.com/wp-dyn/content/article/2008/12/04/AR2008120403537_pf.html.
9. http://www.bmj.com/content/337/bmj.a2338.
10. Stein, "Happiness Can Spread."
11. Nataly Kogan, cofounder of Happier, interview with author, March 22, 2013.

Chapter Seventeen

1. Bruce Sterling, "State of the World 2009," topic 343, remarks at Inkwell: Authors and Artists Conference. http://www.well.com/conf/inkwell.vue/topics/343/Bruce-Sterling-State-of-the-Worl-page01.html.
2. http://www.investopedia.com/terms/e/economics.asp.
3. John C. Havens, "The Value of a Happiness Economy," *Mashable,* June 13, 2012, http://mashable.com/2012/06/13/happiness-economy/.

Chapter Eighteen

1. Robert F. Kennedy, "Remarks at the University of Kansas" (speech, University of Kansas, Lawrence, KS, March 18, 1968), http://www.jfklibrary.org/Research/Research-Aids/Ready-Reference/RFK-Speeches/Remarks-of-Robert-F-Kennedy-at-the-University-of-Kansas-March-18-1968.aspx.
2. Ibid.
3. "Gross National Happiness," *Wikipedia,* accessed 09:43, July 26, 2013, http://en.wikipedia.org/wiki/Gross_national_happiness.
4. A. H. Maslow, "A Theory of Human Motivation," *Psychological Review,* 50 (1943): 370–96, http://psychclassics.yorku.ca/Maslow/motivation.htm.
5. Daniel Kahneman, Alan B. Krueger, David A. Schkade, Norbert Schwarz, and Arthur A. Stone, "A Survey Method for Characterizing Daily Life Experience: The Day Reconstruction Method," *Science* 306, no. 5702 (December 3, 2004): 1776–80, http://www.sfu.ca/~kathleea/phil100/lectures/100_08.Last.pdf.
6. N. Lathia, K. Rachuri, C. Mascolo, and P. Rentfrow, "Contextual Dissonance: Design Bias in Sensor-Based Experience Sampling Methods," paper presented at ACM International Joint Conference on Pervasive and Ubiquitous Computing, Zurich, Switzerland, September 8–12, 2013, http://www.cl.cam.ac.uk/~nkl25/publications/papers/lathia_ubicomp13.pdf.
7. Daniel Kahneman, "The Riddle of Experience vs. Memory," TED Talk video, February 2010, http://www.ted.com/talks/daniel_kahneman_the_riddle_of_experience_vs_memory.html.
8. Clifford Cobb, Ted Halstead, and Jonathan Rowe, "If the GDP Is Up, Why Is America Down?" *Atlantic,* October 1995, http://www.theatlantic.com/past/politics/ecbig/gdp.htm.
9. Jon Hall, head of National Human Development Reports Unit, United Nations; interview with author, July 17, 2013.
10. Ibid.

11. Ibid.
12. http://wwbp.org/about.html.
13. H. Andrew Schwartz, Johannes C. Eichstaedt, Margaret L. Kern, Lukasz Dziurzynski, Megha Agrawal, Gregory J. Park, Shrinidhi K. Lakshmikanth, Sneha Jha, Martin E. P. Seligman, and Lyle Ungar, "Characterizing Geographic Variation in Well-Being Using Tweets," paper presented at Seventh International AAAI Conference on Weblogs and Social Media, Boston, 2013, http://wwbp.org/papers/icwsm2013_cnty-wb.pdf.
14. Jon Hall interview.
15. Timothy W. Ryback, "The UN Happiness Project," *New York Times*, March 28, 2012, http://www.nytimes.com/2012/03/29/opinion/the-un-happiness-project.html?pagewanted=all.
16. H. E. Mr. Jigmi Y. Thinley, "Address by the Hon'ble Prime Minister on Wellbeing and Happiness" (speech, UN Headquarters, New York, NY, April 2, 2012), http://www.cabinet.gov.bt/?p=737.
17. Mark Anielski, *The Economics of Happiness: Building Genuine Wealth* (Gabriola Island, BC: New Society Publishers, 2007), xvii.
18. Mark Anielski, president and CEO of Anielski Management, Edmonton, Alberta; interview with author, July 22, 2013.
19. Ibid.
20. Jon Hall interview with author.
21. Enrico Giovannini, minister of labor and social policies, Italy; interview with author, August 1, 2013.
22. Ibid.
23. Ibid.
24. John Tierney, "How Happy Are You? A Census Wants to Know," *New York Times*, April 30, 2011, http://www.nytimes.com/2011/05/01/us/01happiness.html?pagewanted=all&_r=0&gwh=9FE440DFF7941255517B3A62AF94AB9F.
25. Daniel Hadley, director of Somerstat, interview with author, June 7, 2013.
26. Ibid.

Chapter Nineteen

1. Robin Barooah and Alexandra Carmichael, "Get Your Mood On: Part One," Quantifed Self website, December 9, 2012, http://quantifiedself.com/2012/12/get-your-mood-on-part-1/.
2. http://measuredme.com/about-measured-me/.
3. http://www.slideshare.net/modernmetrix/hacking-happiness-16684885.
4. Konstantin Augemberg, statistician, interview with author, May 9, 2013.
5. Della van Heyst, cofounder of the Billion People Project, interview with author, May 22, 2013.
6. Carrie van Heyst, cofounder of the Billion People Project, interview with author, May 22, 2013.
7. Della van Heyst interview.
8. http://www.upworthy.com/about.
9. http://www.upworthy.com/internet-calls-fat-girl-fat-and-her-response-is-perfect.
10. Arianna Huffington, "Taking the Third Metric Abroad: Redefining Success Goes Global," *Chicago Tribune*, August 1, 2013, http://www.chicagotribune.com/sns-201308010930—tms—ahuffcoltq—m-a20130801-20130801,0,2126559.column.

INDEX